9/9/14

Private Warriors

Washington on $10 Million a Day:
How Lobbyists Plunder the Nation

THE RADIOACTIVE

BOY SCOUT

KEN SILVERSTEIN

Random House · NEW YORK

THE RADIOACTIVE
BOY SCOUT

*The True Story of a Boy
and His Backyard Nuclear Reactor*

Grateful acknowledgment is made to Tom Lehrer for permission to
reprint lyrics from "Be Prepared," copyright © 1953 by Tom Lehrer,
and lyrics from "Wehrner von Braun," copyright © 1965 by Tom Lehrer.
Reprinted by permission.

A portion of this book originally appeared in *Harper's Magazine* in 1998.

LIBRARY OF CONGRESS CATALOGING-IN-PUBLICATION DATA
Silverstein, Ken.
The radioactive boy scout: the true story of a boy and his backyard
nuclear reactor / Ken Silverstein.
p.   cm.
ISBN 0-375-50351-x
1. Hahn, David, 1976–   2. Gifted boys—United States—Biography.
3. Problem youth—United States—Biography.   4. Breeder reactors.
5. Boy Scouts of America.   I. Title.
TK9014.H34S54 2004
621.48'3'092—dc21   2003054811

Printed in the United States of America on acid-free paper

Random House website address: www.atrandom.com

9 8 7 6 5 4 3 2 1

First Edition

Book design by Judith Stagnitto Abbate/Abbate Design

TO MY PARENTS

ACKNOWLEDGMENTS

First and foremost, my thanks to this work's primary editor, Courtney Hodell, who called the day my story on David Hahn appeared in *Harper's Magazine* and said she was determined to see that I turn it into a book. She then patiently and skillfully helped transform a very rough first draft into this final product. Any of the book's shortcomings are surely mine and not hers.

Special thanks also to my agent Melanie Jackson and to several other editors who worked on this book at Random House and Fourth Estate, especially Katie Hall, but also Tim Farrell and Clive Priddle.

I'd also like to thank Clara Jeffery, who edited the *Harper's* story on David Hahn, and Lewis Lapham, that magazine's editor, who commissioned it.

My thanks also to David Hahn and his family for enduring my frequent visits to the Detroit area as well as my endless phone calls asking for just one more piece of information.

Last but not least, thanks to my wife, Clara Rivera, and my two children, Sophia and Gabriel, for their patience throughout.

CONTENTS

PROLOGUE

Men in White: The Nuclear Age
Comes to Golf Manor

There is hardly a boy or a girl alive who is not keenly
interested in finding out about things. And that's exactly
what chemistry is: Finding out about things—finding out
what things are made of and what changes they undergo.
What things? Any thing! Every thing!

—THE GOLDEN BOOK OF CHEMISTRY
EXPERIMENTS, 1960

olf Manor is the kind of place where nothing unusual is
supposed to happen, the kind of place in which people
live precisely because it is more than twenty-five miles from
downtown Detroit and theoretically away from the poverty,
crime, and other complications attendant on that city. It's the
kind of place where money buys a bit more land than in the city,

perhaps a second bathroom, and so reassures residents that they're safely in the bosom of the middle class.

Nestled in a small community called Commerce Township, Golf Manor takes its name from the eighteen-hole golf course at its entrance. The houses wind along about a dozen streets lined with trees both old and varied enough to make Golf Manor feel more like a neighborhood than a subdivision, and the few features that do convey *subdivision*—a sign at the entrance saying, "We have many children but none to spare. Please drive carefully"—have a certain 1950s charm.

A typical summer afternoon in Golf Manor brings, under sunny blue skies and a light breeze, the sight of mothers exchanging gossip as they push their newborns in strollers and the sounds of a few lawn mowers and an occasional whoop from kids jumping on a trampoline or swimming in a backyard pool. In short, it is the kind of place where the only thing lurking around the corner is likely to be a Mister Softee ice-cream truck or a religious devotee peddling the Gospel door-to-door.

However, June 26, 1995, would not go down in Golf Manor's collective memory as a typical day, a fact Dottie Pease grasped immediately when she turned onto Pinto Drive. Dottie had risen early for work and now, with the sun beginning its slow late-afternoon descent, wanted nothing more than to pull off her shoes and drop to the living-room couch for a rest before dinner. As she neared her home, though, Dottie registered a scene so strange that she might as well have driven straight into an *X-Files* episode.

For what she saw, through a fog of bewilderment and dread, was eleven men swarming across her carefully manicured lawn—

three of them enveloped in white, ventilated moon suits. The
attention of the men seemed to be focused on the backyard of the
house next door to hers, specifically on a large, wooden potting
shed that abutted the chain-link fence between the properties.
Thick, leafy branches from a tree in Dottie's yard covered the top
of the shed, which sat directly in front of a swimming pool that
was largely empty save for a small forest of six-foot weeds
growing up through cracks in the bottom.

The three men in protective suits were dismantling the
potting shed with electric saws. Several others were running an
industrial-strength vacuum cleaner across the grass and through
the swimming pool. The crews were dumping the shed remnants
and vacuumed debris into large, jet-black steel drums
emblazoned with bright yellow signs warning of radioactivity.

Dottie parked her car in her driveway and ran to join a group
of about twenty Golf Manor denizens who were mustered in
front of the house where the men were working. The moods on
the block ranged from perplexed and anxious to flat-out
panicked. The fear only grew as the chainsaw-wielding clean-up
crew, who Dottie learned worked for the federal Environmental
Protection Agency, offered no explanation for its activities
beyond vague and empty-sounding assurances that there was
nothing for residents to worry about.

In lieu of information, Golf Manor residents had only
speculation. Unfortunately, they had little to go on. The home
that so preoccupied the EPA belonged to a middle-aged couple,
Michael Polasek and Patty Hahn, who on weekends were joined
by Patty's teenage son, David. A polite young man with blond
hair, hazel-green eyes, and a freckled face, David didn't seem

the sort to cause serious trouble. But he was not in Golf Manor on the day of the EPA's arrival to answer for himself.

Michael and Patty, on the other hand, did have a few quirks. For starters, they were drinkers whose arguments had on occasion become loud, nasty, and public. Furthermore, Patty, though generally pleasant, had an edgy, hostile streak that could show itself without the slightest warning. Michael seemed less volatile, but he sometimes startled and annoyed neighbors by behaving like an overgrown adolescent. One of his occasional pastimes was relaxing in the backyard with a beer and a small pile of M-80s (a powerful firework whose kick is equivalent to a quarter stick of dynamite), which he detonated in the abandoned swimming pool.

Still, Dottie had never spotted anything seriously weird at the house next door—at least not weird enough to explain the federal intervention unfolding before her eyes. Now, though, as she huddled with her neighbors and tried to make sense of the situation, Dottie heard one resident who claimed to have woken late on a recent night and seen, from a back window, the potting shed emitting an eerie glow. This pushed Dottie's alarm to new heights, she later recalled. "I went inside and called my husband. I said, 'Da-a-ve, there are men in funny suits walking around out here. You've got to do something.' "

Bill Larson, a newcomer to the neighborhood, lived just three doors down from Michael and Patty. He was in an especially fine mood when he pulled onto Pinto Drive that afternoon, as he had just been visiting his wife and three-day-old son at the hospital. But as he daydreamed about bringing his newborn home to this cozy nook of a neighborhood, he was suddenly confronted with the EPA spectacle. Great, he thought, I've got a new house for my family and it's sitting in the middle of a toxic dump.

For the next three days, the EPA crew worked away in Michael and Patty's backyard. They broke down the potting shed until nothing remained in its place but a patch of dirt. They painstakingly vacuumed up every last piece of debris and dust from the work site. And then they loaded the steel drums with the radioactive-warning signs onto a flatbed truck, which departed for points unknown.

All the while, the workers refused to provide residents—or the handful of TV and newspaper journalists who dropped by after hearing about the story from Dave Pease—with concrete information about their assignment. A mixture of good-neighborliness and timidity kept residents from questioning Patty or Michael, who didn't make themselves available for inquiries in any case. During the entire ordeal, Patty emerged just once and very briefly, trying angrily to shoo away onlookers. Otherwise, she and Michael were sequestered in their home, with the phone off the hook.

In fact, neither the Peases nor the Larsons nor anyone else in Golf Manor discovered precisely why the EPA had briefly invaded their neighborhood. When asked for details some years later, several residents mumbled something about a small-scale chemical spill. No, others weighed in, the potting shed had sat atop the remains of an industrial-waste site created decades earlier. Still others recalled having heard—probably from the few brief stories filed by reporters, who were hindered by official stonewalling—that young David Hahn was involved and that it had had something to do with a small amount of mildly radioactive material.

This last group came closest to the truth, but they knew only a small part of the story. If Golf Manorites had been fully informed at the time of the EPA cleanup, their concern and panic might well have given way to full-blown hysteria. For federal authorities had set up camp in their neighborhood because they'd discovered levels of radiation in the potting shed so high as to place the area's forty thousand residents at serious health risk. Calm, placid Pinto Drive had in fact been the epicenter of a surreal crisis that had triggered the government's Federal Radiological Emergency Response Plan, the protocol for dealing with any public exposure to radiation, be it a small-scale radioactive release from an industrial accident or a major crisis, such as the near meltdown of the Three Mile Island nuclear-power plant in 1979. Indeed, the EPA's precision strike on the potting shed had been planned after consultations with other government agencies, including the Nuclear Regulatory Commission and the Federal Bureau of Investigation.

Most bizarre of all, David Hahn had done far more than play around with a few radioactive elements. He'd attempted to build a model breeder reactor in his backyard, an effort that grew out of his quest to win an atomic-energy merit badge from his local Boy Scout troop.

THE RADIOACTIVE

BOY SCOUT

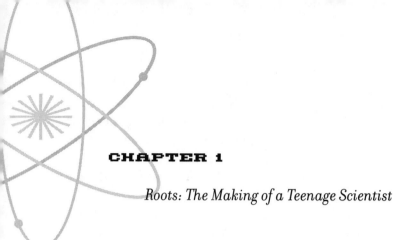

CHAPTER 1

Roots: The Making of a Teenage Scientist

You—Scientist!

—THE GOLDEN BOOK OF CHEMISTRY
EXPERIMENTS, 1960

avid Hahn's earliest memory seems appropriate in light of later events; it is of conducting an experiment in the bathroom when he was perhaps four years old. With his father at work and his unmindful mother listening to music in the living room of the family's small apartment in suburban Detroit, he rummaged through the medicine chest and undersink cabinet and gathered toothpaste, soap, medicines, cold cream, nail polish remover, and rubbing alcohol. He mixed everything in a metal bowl and stirred in the contents of an ashtray used by his mother, a chain-smoker. "I was trying to get a magical reaction, to create something new," he remembered later. "I thought that the more things I threw in, the stronger the reaction I'd get."

After he finished blending the ingredients together, young David was disappointed to see that all he had in the bowl was a lifeless, grayish glob. Hence, he went back to the cabinet beneath the sink and pulled out a bright-blue bottle, which years later he realized was probably a drain-cleaning product. He uncapped the bottle and poured a healthy amount into the bowl; soon, the mixture began to bubble and threatened to boil over. In a panic, David flushed the contents of the bowl down the toilet. His parents never knew what happened, and David promised himself that he would never again try something so foolish. It was the first of many similar vows made over the years, all broken in short order. It also established a pattern: experiment, trouble, cover-up.

If David was a slightly odd child, his parents, lost in their own preoccupations, hardly noticed. His father, Ken Hahn, grew up in the Detroit area along with his four brothers and sisters. Ken's father was a skilled tradesman, a tool-and-die maker who worked for General Electric and Pratt & Whitney. At night, Ken would sit with his dad and pore over blueprints of the tools his dad made during his workday. By the time he reached Henry Ford High School, Ken had decided to pursue a similar career, though he was fascinated by the idea of drawing the blueprints, not building the tools. He enrolled in a college-prep program for mechanical engineering and after graduating attended Lawrence Technological University, a local school.

Ken was so wrapped up with his engineering studies that he had little time for dating or romance. But while a sophomore at

Lawrence Tech, he and a friend were cruising Woodward Avenue just outside of Detroit when they spotted two pretty girls driving alongside his Chevy Chevelle. After signaling for them to pull into a Big Boy hamburger drive-in, Ken zeroed in on nineteen-year-old Patty Spaulding and came away with her phone number. For Ken, it was love at first sight. "She was cute as a bug," he remembered later, proudly showing off a picture of a beautiful young woman with a bouncy smile.

But Patty, having recently ended a stormy relationship, was initially aloof. She had not had many positive experiences with men. Patty had been raised in a poor region of West Virginia, and her father had abandoned the family when she was young. Her mother, Lucille, had packed up and moved the family to Detroit, where they had relatives. Lucille found work at a doctor's office, and the family moved into the middle class, albeit at the lower end of that category. It wasn't an easy life, but it was better than West Virginia.

Ken was a determined suitor, though. After a four-year courtship during which he displayed the same tenacity that he normally reserved for work-related engineering challenges, Ken finally wore down Patty's resistance. They were married in July 1974.

Like those of all residents of contemporary Detroit, Ken and Patty's lives were shaped physically, economically, and socially by the automobile industry. The metropolitan area was then home to Ford, General Motors, and Chrysler, as well as to thousands of small shops that produced machine parts, brake linings, and industrial tools for the Big Three automakers. Soon after the wedding, Ken found a job as a mechanical engineer at a

General Motors subcontractor, and he and Patty moved into a suburban apartment complex not far from his office. David, their only child, was born on October 30, 1976.

Ken worked long hours, designing robotic welding machines and other assembly-line equipment. He left home punctually at six in the morning, rarely returning before six in the evening and sometimes not until after David had gone to bed. Tightly wound, Ken was a dutiful husband and father but not a demonstrative one. Combined with his constant air of preoccupation, his reserve must have been confounding to a child. Even when Ken was around the house, there was little interaction between father—David remembered him as "always off in a fog"—and son, who developed an especially close bond with his mother.

In contrast to her husband, Patty was outgoing and affectionate. She loved children and painted watercolors of kids at play, some which were displayed for years at the Detroit Children's Hospital. Patty lacked Ken's focus, though, and had a hard time sticking with anything. She'd dropped out of high school three weeks prior to graduation and, despite several attempts, never got around to completing her GED. For a time, she talked about becoming a model and even put together a portfolio before abruptly abandoning the idea.

Patty doted on her son and gave him the attention he couldn't get from his anxious and distant father. When David wanted a basketball hoop in his room, Patty made Ken put one up. If David liked a song, she'd play it for him over and over again. As David remembered, "My mom might be sleeping in her room when I got home from [elementary] school, but she

always popped up to see me, and we'd do my homework together.
If I did a drawing at school, she always put it up on the wall and
bragged about how great it was to whoever came over, even the
plumber. I thought she was the most wonderful person in the
world."

But troubles began to dog Patty, though David was largely
unaware of what was happening. She developed the drinking
problem that ran in her family. A few years after David was born,
she began to hear voices and thought strangers were after her.
She was diagnosed with depression and paranoid schizophrenia.
A variety of antipsychotic medications were prescribed. Fearing
someone was trying to kidnap David, Patty took to changing the
locks on the doors. She heard ghosts in the apartment building
and would take David by the hand, creep down the basement
stairs with a flashlight, and make sure nothing was lurking there.
Ken hired a retired woman who lived nearby to check up on Patty
and his son when he was at work, but by the time David was four
Patty's condition had deteriorated so badly that she had to be
committed to a mental hospital.

To explain her absence, Ken told David that his mother had
been hurt when her car skidded off the road during a rainstorm.
David suspected the story wasn't true—it couldn't have provided
much comfort in any case—and felt completely abandoned. Upon
hearing that Patty would have to "be away for a while," he hid
behind the couch in the living room, clasped his knees to his
chin, and rocked himself back and forth.

Patty returned home six months later, and though she wasn't
hospitalized again after her release her illness lingered and
deepened. She rarely worked and spent most of her time at the

apartment, caring for David when he wasn't at school and watching TV, listening to Top 40 hits, and playing cards with her girlfriends when he was. Though Patty still pampered David, she became somewhat less attentive. Left on his own, David developed a wild imagination. He built elaborate sets in his room—caves built from pillows and forts constructed in his closet—on which he could act out games with make-believe space explorers and superheroes. He fantasized endlessly about comic-book hero Spider-Man, the alias of Peter Parker, a dweebish, bespectacled high school student who gained superpowers after being bitten by a radioactive spider.

Meanwhile, the marriage between David's parents was falling apart, riven by financial troubles and Ken's frustration with Patty's failure to look for work or, in his view, deal with her mental troubles. As David peered out from his bedroom, his parents would scream at each other across the living room, and on occasion Patty would hurl a vase or a lamp at the wall. In 1985, when David was nine years old, his parents finally split, and Patty lost custody of her son. It was then that David's troubles really began.

David stayed with his father, who soon began dating a GM engineer named Kathy Missig. Ken and Kathy—whose daughter from a previous marriage, Kristina, was David's elder by a year—didn't marry until six years later, but within a year of meeting they bought a house together in Clinton Township, a conservative working-class area about twenty miles north of downtown Detroit.

Thanks to Detroit's devotion to the automobile, urban planning and mass transit were, and are, almost unknown to the

region. Clinton Township, like other outlying areas, was an
endless sprawl of fast-food restaurants, strip malls, shopping
centers, and other signposts of suburbia. The Hahns' new home
was a small but cozy split-level. The family room boasted birch
paneling and a fireplace, while David's bedroom, on the top
floor, looked out on a diamond-shaped deck in the backyard,
with the requisite affordable luxuries of a barbecue grill, patio
furniture, and an aboveground swimming pool.

Ken remained wrapped up with his job and was rarely at
home and even more rarely available to his son. He'd often get
back long past the dinner hour, so Kathy would leave a plate of
food warming for him in the oven. David saw his dad as a hard
worker but conservative and living a boring lifestyle. "He talked
a lot about work and people I didn't know anything about," he
said. "He was always telling me that he didn't spend much
money, just a few dollars a day. I wanted my life to be more
exciting than that."

Nor was David close to Kathy, who had the impossible task of
replacing David's adored mother. He resented her efforts to
impose rules and regulations in his father's absence and felt that
Kathy favored Kristina, with whom David had a strained
relationship. One Christmas Day, David remembered with pain
many years later, Kristina excitedly opened a number of
carefully wrapped presents while he received only a "mad
scientist" kit containing multicolored goop, something he
thought was for babies. For David, growing up was a battle, and
he felt that he had to fight for everything, be it affection or
money. "With the new family, no one paid attention to what I was
doing," he said. "The caring went away."

Weekends and holidays offered a certain relief, as David stayed with his mother and her new boyfriend, Michael Polasek. They lived about forty-five minutes away in Commerce Township, which exuded the same comfortable—if slightly less affluent—suburban atmosphere as Clinton Township. An amiable but hard-drinking retired GM forklift operator with a ninth-grade education, Michael looked like he had stepped out of a casting call for *Grease*. He collected muscle cars, like a yellow 1968 Camaro with black racing stripes and a cherry-red 1983 Jaguar, which sat, buffed and shined, in the driveway. Michael was just as finicky about the house, which he kept spotless even as he fawned over their five cats.

Michael loved to read, especially books about American history, which made a big impression on David. A further and very different attraction was his ready supply of cherry bombs and bottle rockets. Michael, Patty, and David often drove out into rural Michigan and hiked along trails that went miles and miles into the wilderness. After pitching a tent and building a campfire, Michael and David would set off a cache of fireworks as Patty flipped burgers on the grill.

Life in Commerce Township, though, was far from idyllic, especially given Patty's continual battle with alcoholism and mental decline. While Patty desperately tried to hold herself together when David was around, she suffered violent mood swings and several times disappeared for days. Pictures of Patty from these years bear no resemblance to the sparkling girl in Ken's photograph. In place of the ebullient teenager was a sullen woman, overweight and bloated from drinking.

Despite the fact that David was shuffled between eccentric

and erratic households, his early years seemed ordinary and
happy enough. A short, good-looking kid with an angelic smile,
he joined baseball and soccer leagues, explored the
neighborhood and woods near his father's home, and spent
hours with his friends riding bikes, shooting hoops at the school
playground, and playing in a fort that they built out of two-by-
fours in the branches of an oak tree near a forested trail at the
edge of the subdivision.

There were, however, some early indications of a quirky
personality. David began exhibiting a mildly destructive bent
around the age of five—surely an angry reaction to his mom's
hospitalization. He let air out of tires on cars parked in the
neighborhood. Once, with a group of friends, he started a fire in
the woods that escaped control and required the fire department
to put out. (David and his buddies fled the scene before the fire
trucks arrived and never told anyone they were responsible for
the small blaze.)

Just a few years later, David became utterly fascinated with
everything mechanical. Ken bought him an endless supply of
robots, model kits, radios, tape recorders, printers, even simple
remote controls. David didn't want the items for their intended
purposes but instead spent hours disassembling and
reassembling them to try to figure out how they worked. "He
didn't want the plastic body," his father remembered. "He
wanted the guts." David himself says he was intrigued with the
idea of using parts from everyday products to make something
entirely new. He passed hours dreaming about building a jetpack
he could strap to his back and use to fly over the neighborhood
or a robot that could mow the lawn or take out the garbage.

David spent a good deal of time with Patty's mom, Lucille, who lived in the town of Berkley, four miles outside the Detroit city limits. Lucille remembered the young David as a hyperactive, skinny kid surrounded by huge piles of batteries, which he'd somehow string together and use as a power source for lightbulbs and small appliances. "He spent a lot of time trying to build things," said Lucille, a short woman whose height was extended by a bouffant hairdo. "He didn't play outside as much as most kids do."

David's budding scientific persona bloomed especially brightly in Golf Manor, largely because he had little responsible adult supervision. Like many young boys, David soon discovered the existential pleasures in blowing things up, a pursuit to which he devoted considerable time and energy. He perfected a recipe for homemade gunpowder and manufactured small bombs, which he and Michael, his partner in demolition, detonated in the woods, at the bottom of their empty swimming pool, and anywhere else that the urge struck. "We went out to the boondocks and blew those things up," Michael, who even as a middle-aged man loved a big boom, recalled with a laugh and an admiring shake of his head. "He scared the hell out of me."

At first glance, David's behavior might not appear terribly different from the mix of adventurousness, curiosity, and mischief displayed by plenty of young boys. In fact, until the age of ten, David's scientific fantasies (and explosives workshop) were merely distractions; sports, skateboarding, and exploring the woods with his friends were the focus of his life. The first sign that a hobby was becoming a mania came when John Sims, Kathy's father, gave him a used, long out-of-print copy of *The*

Golden Book of Chemistry Experiments. Just about every young boy at some point goes through a chemistry phase, which typically doesn't advance far beyond mixing a few chemicals in a test tube to produce smoke. In David's case, the *Golden Book* provoked a more lasting and powerful reaction, akin to a chemical pairing of nitric acid with glycerin.

Written by Robert Brent and published in 1960, the *Golden Book* practically vibrated with a relentlessly upbeat tone and colorful futuristic illustrations. One showed a happy, maskless farmworker spraying chickens with a cloud of insecticide, while another featured three workers operating a spotlessly clean nuclear reactor, a technology fairly new at the time the *Golden Book* was published.

Comparable chemistry books sold today are designed for parents as much as for kids, offering the wan pleasures of experiments that require no glass pieces and no open flames and use only environmentally safe materials. Take *The Usborne Book of Science Experiments* (1991), widely available in toy stores. Among the experiments described in its pages are watching a sunset in a box, making a garden in a jar, observing microbes under a microscope ("Yeast comes to life!"), and constructing a wormery. To gauge by the *Usborne Book*'s cautious tone, science is something to be approached with trepidation more than excitement. Warning signs abound throughout the text: Treat the worms in your experiments gently, and put them back where you found them; wash your hands thoroughly before handling foodstuffs; never use household electricity for experiments.

The *Golden Book*, by contrast, promised to open the doors to a brave new world. It was the era of JFK and the New Frontier, of satellite launches and the race to the moon. The sky truly was the limit. "Chemistry is one of the most important of all sciences for human welfare," the *Golden Book* asserted with unwavering optimism. "Chemistry means the difference between poverty and starvation and the abundant life."

And this was just the start. New and improved chemicals, David learned, would make it possible "to keep food fresh without refrigeration in any climate." Future travel would occur largely at supersonic speeds. "Planes and rockets will require materials that can stand tremendous heat and new fuels capable of providing enormous energy. Chemistry will provide them."

Of all the wonders of the future described in the *Golden Book*, few promised more "for the welfare of all humanity" than the harnessed powers of the atom. Nuclear power, he read, would be the driving force of a scientific revolution that would soon transform the planet as the "force hidden in the atom will be turned into light and heat and power for everyday uses."

This especially intrigued David because he thought nuclear power might be a solution to his family's problems. The energy crisis of the 1970s and early 1980s had hit Detroit—the nation's biggest car producer and a major manufacturing hub—especially hard. The soaring cost of energy was a frequent topic of conversation in the Hahn household—David heard over and over again that the United States was dependent on foreign countries for oil—and the growing monthly electric bill was a constant source of conflict between Ken and Kathy.

Perhaps nothing David read in the *Golden Book* affected him more than the story of Marie and Pierre Curie, whom he worshiped with the same intensity with which his friends revered basketball great Earvin "Magic" Johnson, who had played at Michigan State University in nearby Lansing a few years earlier and was now a superstar with the Los Angeles Lakers. The Curies were the heroes of Chapter 10 of the *Golden Book*, which featured a picture of the couple working diligently in their lab, bathed in the soft, embracing glow of radium emanating from a small jar.

The Curies spent years investigating a mineral called pitchblende, which emitted rays that were stronger than its uranium content could explain. To discover the source of the rays, the couple obtained eight tons of ore from Bohemia, boiled it down with acids, and purified the remains with processes they invented. After two years of backbreaking labor in their drafty shed, the Curies isolated and removed the uranium. They soon discovered that the substance that remained glowed powerfully in the dark; the Curies named this new substance radium, a name derived from the Latin *radius,* meaning *ray.* (In Roman times, a radius was a bright, long-lasting candle used to illuminate stadiums hosting gladiatorial events.) Marie Curie named a second element found in the pitchblende polonium, after her native Poland.

The story of the Curies, said the *Golden Book,* demonstrated "all the features that show the nature of the true scientist. Curiosity first."

Written in an era well before lawyers began earning such good livings off the proponents of bad advice, the *Golden Book* today seems amazingly oblivious to the volatility of the experiments it described. One chapter, entitled "Chlorine— Friend and Foe," carried an illustration of a soldier dashing into combat wearing a gas mask. The good news here was that chlorine had a number of beneficial uses, not least that it protected the nation's drinking-water supply. On the downside, chlorine was "dangerous when used improperly because it affects the lungs. As a 'poison gas' it caused many casualties in World War I." Tens of thousands of casualties to be precise— soldiers whose lungs filled with a greenish vapor that stripped away the mucus lining, causing them to drown in their own body fluids.

Having passed briefly over the deadly legacy of chlorine gas, the *Golden Book* went on breezily to teach readers how to brew their own in a home lab. The three-step recipe was simple, requiring as equipment only three widemouthed bottles corked with rubber stoppers and connected with glass tubes. In step one, young scientists combined a touch of Sani-Flush and Clorox in bottle A, whereupon chlorine gas would form and fill bottle B, with the excess of gas then being absorbed by a lye-water solution in bottle C.

The book warned somewhat inadequately that when removing bottle B, bottles A and C should be connected to prevent chlorine gas from "getting out in the room." Furthermore, readers should perform the experiments outdoors or before an open window. Most important of all, the book cautioned, "Be careful not to breathe fumes!"

David had always had an obsessive streak. Before discovering the *Golden Book*, he had built a huge baseball-card collection, which he relentlessly organized and reorganized, arranging the players first in the order of their batting averages, then their number of career home runs, and on and on. Now his punctiliousness turned to science.

Following the detailed instructions in the *Golden Book*, David set up a laboratory at his father's house. His first base of operations was his small bedroom. He bought beakers, Bunsen burners, and test tubes, as well as sodium hydroxide, potassium nitrate, sodium nitrate, ammonia, sulfur, magnesium, and a vast range of other chemicals and raw materials. David frequently had up to a dozen experiments and projects in progress simultaneously. In the early days, these were simple matters, such as assembling clocks and alarm systems from kits. He also conducted most of the two hundred experiments described in the pages of the *Golden Book*, from learning simple evaporation and filtration techniques to making rayon and alcohol.

Having imbibed the *Golden Book*'s worshipfully uncritical approach to science and braced by every teenager's assumption of invulnerability, David gave little thought to any potential dangers from his experiments. He produced chloroform, a primitive anesthetic commonly used during surgery in the mid-1800s, from a recipe in the *Golden Book* that involved gently heating a mixture of ethanol and sodium-hydrochloride solution. The book urged readers to sniff carefully and savor the "peculiar sweetish odor of chloroform"; David took the challenge but apparently sniffed a bit too vigorously and ended up flat on his back. He estimated he was out for more than an hour.

His parents, charmed by David's interest in what seemed like a stimulating, safe pastime—he carefully kept from them episodes such as his chloroform blackout—initially offered encouragement and modest financial support. But science, as practiced by David, wasn't cheap. To pay for his activities, he started the Big D Lawn Mowing Service, posting handmade flyers around the neighborhood, passing them out at community events, and slipping them under doors. Business boomed; when he wasn't in his bedroom lab, David was pushing the family mower over the neighbors' lawns.

David's neighborhood friends would stop by to see if he wanted to throw around a Frisbee, hang out at a park, shoot some hoops, or just sit around and watch TV. He now declined most invitations and headed for the public library, where he'd comb shelves and pester the reference librarian in a quest for information on his new passion. By the age of twelve, David had become a voracious reader of scientific books, including his father's college chemistry books, which he digested without difficulty. That impressed Ken, though he seemed to regard it as a quirk, not a sign of precocious scientific talent. When David spent the night at Golf Manor, his mother would wake to find him asleep on the living-room floor, surrounded by open volumes of the *Encyclopaedia Britannica.* Frustrated by his inability to memorize material he deemed important to his unfolding scientific career—like the entire *Merck Index,* a reference guide that catalogues thousands of pharmaceuticals, compounds, and hormones—David sought to increase his brainpower with megadoses of two products he bought by mail order: ginkgo biloba, an herbal remedy, and a steroid called pregnolone.

Now in middle school, David had a hard time concentrating

on anything but science. Never the most popular kid, he gained a reputation as a space cadet as he walked the halls of the school reading books about nuclear power and chemistry, excitedly talking to himself when he had an epiphany. Even worse, David experimented on himself with hair dyes he had concocted. The dyes were as strong as store-bought products—too strong, in fact. Once, he was unable to rinse out the various shades he had manufactured over the weekend and turned up at school Monday morning with his blond hair turned green, brown, and black. During English class, a group of girls teased him mercilessly and fired spitballs at him every time the teacher looked away.

Adults, too, began to treat David as something of a freak. He once told a science teacher that he was studying radioactive materials and wanted to understand the difference between two isotopes. The teacher shook his head, told him to quit making up stories, and walked away.

One of the few people David could talk to about his growing interest in science was John Sims, Kathy's father. John lived in the nearby town of Mount Clemens, and whenever he dropped by the house he would inevitably find David sequestered in his bedroom amid a jumble of electrical wires, batteries, and test tubes. A retired GM engineer with a strong academic background in chemistry, John took David's experimenting seriously and didn't treat him like a flaky kid. John was impressed but startled by David's quick mastery of scientific material. "He was a very inquisitive boy and was experimenting with things he didn't realize the full consequences of," his stepgrandfather later said. "Within a few years he progressed well beyond my level of chemistry."

The kingdom of science offered David a welcome refuge from many traumas: a work-obsessed father, a mother with crippling mental problems, the stress of two new homes and two new stepparents, not to mention the normal growing pains of preadolescence. He spent hours locked in his room, poring over pictures and absorbing the texts from a pile of scientific books, while the world outside faded away to the level of background noise. "My family's problems seemed unsolvable, so whenever I was in trouble, or my parents were fighting, I'd go right back into my books," he said.

There was something else about science that appealed to David: He was good at it. Being smaller and skinnier than most boys his age, he'd never excelled at sports and was the sort of kid who got picked last when it came time to choose up sides during gym class. Nor had he shone in the realm of academics, as he was easily distracted and bored in most of the classes. It showed in his report cards, which were a mix of Cs and Ds, with the latter usually in the majority. Mastering the *Golden Book*'s simple experiments gave him a sense of control, self-confidence, and self-respect, and at the same time it transformed his view of the world from a narrow, cramped battleground to a universe whose vast horizons were ever expanding in lockstep with the march of science. "I tried other stuff but I never got anywhere," he said. "Science was something that I could master. Finally, there was something that I had control over."

In *The Making of the Atomic Bomb*, Richard Rhodes referred to several psychological profiles of pioneering American scientists conducted in the early twentieth century. One of the studies, which used the Rorschach and Thematic Apperception tests, found that scientists were frequently raised in homes where the father had died early, worked away from home, or remained so distant and nonsupportive "that their sons scarcely knew them." Those with living fathers described them as "rigid, stern . . . and emotionally reserved," and the scientists themselves were "slow in social development, [and] indifferent to close personal relationships [or] group activities."

A second study cited by Rhodes had produced a "composite portrait" of America's most eminent scientists. They were sons of professional men, typically engineers or doctors. This composite scientist "in boyhood began to do a great deal of reading. He tended to feel lonely and 'different' and to be shy and aloof from his classmates. . . . Once he discovered the pleasures of [science], he never turned back. He is completely satisfied with his chosen vocation. . . . He works hard and devotedly in his laboratory, often seven days a week. . . . Better than any other interest or activity, scientific research seems to meet the inner need of his nature."

David doesn't precisely match this profile—which is a rather long-winded description of what adolescents know more straightforwardly as a geek—but the similarities are abundant and striking. In one important way, though, David greatly differed from the composite scientist: He had no "fatherly science teacher" to guide him through his early years, and hence his experimenting took place with little input—or encouragement or supervision—from adults.

One of David's favorite pastimes was building model-kit rockets, which he souped up with his own fuels and elaborate design and engine innovations. When David was twelve years old, Ken bought him a Helicat rocket from Toys "R" Us. The bright yellow Helicat came with snap-on carbon-dioxide cartridges that served as the propulsion system, giving it a range of about six hundred feet—high enough that safety instructions warned that the rocket should be launched only in a large field away from power lines, tall trees, and low-flying aircraft. David spent hour after hour in the backyard looking for ways to increase the Helicat's range. He experimented with various types of siding to decrease wind resistance before settling on a thin aluminum sheet. Then he devised a means to attach additional carbon-dioxide cartridges to boost the rocket's power. By the time David was finished, the rocket soared at least three times higher than advertised. "That thing would shoot up straight as an arrow; it must have reached almost a mile into the sky," Ken recalled. "He spent so much time working on it that it got to the point where I was sorry I bought it."

David also became expert at making his own fireworks. His formula, which he adapted from a primer on the subject he found at the local library, called for large quantities of magnesium shavings. Fortunately, Michael's son from an earlier marriage, who was nearly twenty years older than David, drilled out manifolds at an auto shop. At David's request, he delivered bags of small, curly magnesium tailings to the house in Golf Manor.

David combined the magnesium with various metals, placed the mixtures in bowls, and set them afire, which produced effects like oversized sparklers. Copper and magnesium made for a green sparkler, while potassium produced a purplish hue. Strontium—which David secured by splitting open highway-safety flares that he bought at a hardware store—created a bright red glow. "I felt like I was learning and accomplishing something with the fireworks and explosives, but it was also a way for me to get attention," David admitted. "I was always a superstar at the Fourth of July. I tried to beat everyone on the block with big bangs and beautiful colors. That was my time."

One of David's chief successes was a power skateboard he built by attaching a small electronic motor to the bottom of a store-bought model. He controlled the speed—which topped out at more than ten miles per hour—with a jury-rigged joystick of his own design. Save for a handlebar rising from the base, David's creation was quite similar to the motorized skateboards that a few years later became immensely popular with teenagers in the United States and Europe.

David used the skateboard to tool around the neighborhood until Kathy took it away after he fell off while cruising at maximum speed and sprained his wrist. This was one of many episodes that resulted in a trip to the medicine chest, the doctor's office, or worse. One particularly scary accident occurred when a propane stove David and a friend were fiddling with suddenly ignited, searing his arm with second-degree burns. "He had to go to the hospital so many times I lost count," his stepmother said, just a touch facetiously. "Finally I'd say, 'Ken, you take him, I'm going to bed.' "

Before long, David was becoming known as Clinton Township's homegrown mad scientist. He was frequently seen wandering the streets of his neighborhood, pockets bulging with copper wire, splice connections, transistors, and a soldering gun. When new friends found out David was her stepbrother,

Kristina would get a flurry of questions about what it was like to live with such a strange guy.

As David's interest in science grew, he withdrew further and became increasingly secretive. The refrigerator at Ken and Kathy's was perennially stuffed with blenders, soup cans, and stainless-steel mixing bowls containing David's chemical mixtures, which he warned other members of the family, in vague though vigorous terms, to leave sealed.

Then there were the smelly concoctions that fair-skinned David manufactured in an effort to protect himself from sunburn. He took stalks of celery—which contains 8-MOP, a substance that activates an enzyme in the skin that increases the production of melanin, a dark pigment—and ran them through a blender until they turned into a thick green sludge. He read in the *Merck Index* that 8-MOP was soluble in acetic acid, so he added vinegar (which is 30 percent acetic acid) to the sludge, heated the mixture on the stove top, and applied it directly to his skin.

In his ongoing efforts to improve his memory and general brainpower, David also blended "shakes" out of green vegetables and Portobello mushrooms, which produced a horrible brownish muck that he believed capable of heightening intelligence. Kristina still remembers her trepidation when opening the door of the refrigerator; she never knew what noxious odor would come wafting out.

With scientific pursuits swiftly becoming the center of his life, David decided to set up a second laboratory at his mother's house in Golf Manor, where he spent weekends and holidays. He found the perfect spot: a long-out-of-use backyard potting shed.

David spent weeks cleaning it up, throwing out old soda bottles, paint cans, copper pipes, and broken-up slabs of concrete, the latter of which he tossed into the empty swimming pool. He finished the job with a coat of white paint and a red radioactive symbol on one wall before finally moving in his lab equipment. The *Golden Book* urged readers to "follow in the footsteps" of the Curies by mastering filtration, evaporation, crystallization, and other laboratory methods they employed. Now David, who had taken the exhortation literally, was ready to do so.

David told Michael and Patty to keep out of the shed so as not to mess up his experiments. In keeping with their laid-back approach to parenting—and perhaps intimidated by his budding confidence and intelligence, especially given their lack of formal schooling—they heeded his demand. He was thus able to expand his research into thrillingly perilous territory. "He was so young, who figured he could do anything really dangerous?" Michael said, years later.

The archetypal suburban American boy with a penchant for chemistry conducts rudimentary gunpowder experiments; David quickly raced past that point. After discovering that it was impossible for a twelve-year-old boy to purchase nitric acid— a powerful chemical with which he planned to make bigger and better explosives, notably nitroglycerin—David attempted to fabricate his own variant from a formula he found in *Modern Chemistry*, one of his father's old college textbooks. (He ignored a warning in the textbook that said that terrible accidents "have occurred through the unintentional discharge of explosives. . . . And under no conditions should an amateur ever try to make an explosive.") David mixed and heated together two commercially available chemicals, saltpeter and sodium bisulfate, then bubbled the gas that was emitted through a container of water to condense it back to liquid.

Like any good scientist, David tested the results. Knowing that nitric acid oxidizes copper, he threw a few coils of copper wire into his product and placed it on a window ledge in the basement of Michael and Patty's home. Before long, the acid was roiling and eating away at the wire, to David's satisfaction. This inventive single-mindedness, so completely absent from the rest of his life, proved to be typical of David's attempts to get hold of whatever he needed for his science projects, no matter how tightly controlled and regulated the materials he was after.

David's most impressive achievement came when he turned a bug zapper from Radio Shack into a remote-control system that increased and decreased the electrical voltage throughout the house. Ken and Kathy were lying in bed ready to doze off one night when lamps on the bedside tables and the TV started going on and off by themselves. Ken raced down the hall to David's room, from which he detected a pungent smell. When he threw the door open, he found his son tinkering with what looked like a galvanized pipe wrapped in copper wire—this was the bug zapper's electrically charged stalk, which David had extracted and wired so that it plugged into his bedroom outlet. The bizarre gadget was emitting sparks that caused the air in the room to crackle with electricity and releasing a stream of smoke that caused Ken's eyes to water. "This was one of the times that I got so mad at him that I [wanted to] chase him down the street," said Ken.

David's invention was as dangerous as it was original; during one test, the modified bug zapper once gave him a shock so severe that it pushed him four feet. But as he did with Patty and Michael, David snowed Ken with sincere promises about a new commitment to safety. Like the parent who believes his kid's story about having eaten something spoiled after he's been discovered stumbling home at midnight and throwing up in the bushes, Ken credulously took David at his word.

Kathy, on the other hand, was becoming exasperated by David. His bedroom, where he carried out most of his experiments, was all but completely destroyed. The walls were badly pockmarked from a multitude of chemical explosions, and the carpet was so stained that it eventually had to be ripped out. Even the padding and plywood subflooring underneath was stained blue from spills of indole, an alkaloid derived from indigo pigment that David used to make natural-highlight shampoos.

Pressed by his wife to do something, Ken finally took action—though, typically, it didn't involve disciplining his son. Instead, he pushed David to enroll in the Boy Scouts, a step that ultimately proved to be as crucial to the unfolding of David's bizarre story as his discovery of the *Golden Book*. A former scout himself, Ken had organized camping trips in his teens and for four years been a patrol leader with Troop 207 in Detroit. He believed that the organization would provide the discipline—and distraction—needed to keep his son out of trouble. He also thought that scouting would give David a chance to bond with other kids and that it might bring father and son together as well. "For me, scouting built character," Ken said. "It taught me leadership skills, how to speak in front of groups, and gave me confidence. I thought it would be good for David and me if he joined."

David was reluctant, thinking that scouting was for nerds, but he grudgingly joined Clinton Township Troop 371. To his surprise, he loved the Boy Scouts. There were camping trips, cookouts, and a host of other outdoor activities, as well as regular meetings where he and other enlistees learned traditional scouting skills, like rubbing sticks together to start a fire and using a compass if lost in the woods. Ken's own father hadn't supported him during his years in scouting, and he was determined to do better. And so while he was otherwise largely absent from David's life, Ken often tagged along to scouting events.

David's scoutmaster, Joe Auito, resembled an aging Deadhead rather than the rock-ribbed conservative of scouting legend. He later recalled David as one of the more active members of the troop—a great swimmer and camper, a fast learner, and a big help with the younger scouts.

But Joe's most vivid memories of David concerned his oddball nature and freelance scientific activities. Once David turned up for a scout meeting with his face tinted beet red and his hair bright orange. As Joe remembers it, David's startling appearance that day resulted from a chemical mishap, though David suspects it was caused by an overdose of canthaxanthin, a steroid that he ordered from the back of a muscle magazine and was taking as part of an experiment on artificial tanning.

One year, David enrolled in a program called Survival Camp, which required scouts to camp outdoors overnight during the dead of winter. Ken had passed through Survival Camp as a young scout, and it made him vastly more appreciative of the comforts of home. He was excited that David would undergo the same experience and hoped that his son would come away with the same feelings after a night out in freezing weather. But David came away with a somewhat different lesson—namely, that science could make life easier and more comfortable, just as the *Golden Book* had promised. He arrived at Survival Camp with a backpack full of batteries, which he rigged up to a heating element that he placed inside his sleeping bag. While most kids shivered through the night and counted the minutes until it was over, David read magazines by flashlight before drifting off for a cozy night's rest.

All the while, David was becoming more and more submerged in scientific reverie. As his intellectual horizons expanded, he began to broaden his reading to more advanced material from the public library. Still, his favorite resource was the *Golden Book*, which he never tired of rereading. At its conclusion, there was a one-page chapter titled "What's Ahead in Chemistry?" "Chemists of the future, working with their brother-scientists, the physicists, will find new ways of harnessing and using the atoms of numerous elements—some of them unknown to the scientists of today," it enticed. "Do you want to share in the making of that astonishing and promising future?"

It was an appeal for converts, and it struck a deep chord with David. Increasingly friendless and estranged from family, he longed to feel himself part of something important, of a larger cause and tribe. With the *Golden Book* as his bible, he joined the atomic fraternity.

CHAPTER 2

From the Radium Craze to the Soaring
Sixties: Science Conquers All

> The physiological action of radium is not unlike a fairy
> tale. It stimulates all cell life, particularly that of the
> enzymes, thus aiding and improving metabolism. . . . No
> toxic or lasting ill effects have been reported.

> —DOUGLAS MORIARTA, RADIUM, 1916

From the beginning, the history of radioactivity and
atomic research has been marked by cycles in which
amazing discoveries produce mass elation only to give way to
fear of the consequences of the new knowledge. The first cycle
began in 1896, when Wilhelm Conrad Roentgen discovered
X rays, a highly penetrating form of electromagnetic radiation

and the first practical application of the radioactive age. Almost immediately, X rays produced astonishing medical breakthroughs, including the ability to diagnose ailments such as tuberculosis and pneumonia, to locate bullets in gunshot wounds, and to kill diphtheria microbes.

As recounted in the delightful book *Nukespeak: Nuclear Language, Visions, and Mindset,* having your own personal X-ray portrait soon became a status symbol in New York, that city always in search of a new fad. There was much discussion about the possibility of seeing through clothing; one firm even offered lead-lined, X-ray-proof underwear. A poem published in a photography magazine in 1896 captured the spirit of the time:

> *The Roentgen Rays,*
> *What is this craze?*
> *The town's ablaze,*
> *For nowadays,*
> *I hear they'll gaze*
> *Thro' cloak and gown—and even stays,*
> *These naughty, naughty Roentgen Rays.*

This wave of excitement persisted despite quickly emerging evidence that exposure to radiation could cause severe injuries and even death. By mid-1897, a year after the discovery of "Roentgen Rays," *The New York Times* was already warning of the dangers of X rays, citing multiple cases of severe burns. By 1905, many doctors who employed X rays had already died or suffered amputations to stave off cancer. Dr. John Hall Edwards, one of the early pioneers, wrote at the time, "I have not experienced a moment's freedom from pain for two years. In cold weather I am unable to dress myself, and the pain experienced cannot be expressed in words."

Even then, researchers and the public gave little thought to the potential dangers of radioactive materials. In fact, a new and more hysterical period of popular, commercial, and scientific enthusiasm was generated by research into radium, which the Curies had discovered in 1898. Sir William Ramsay, a leading expert on the new field of radioactivity, believed there were no limits to what radium might mean to the world. "[The] philosopher's stone will have been discovered, and it is not beyond the bounds of possibility that it may lead to that other goal of the philosophers of the Dark Age—the elixir vitae," he prophesied, referring to the mythical tonic that ensured health and longevity.

According to *Nukespeak*, "The luminous properties of radium soon produced a full-fledged radium craze. A famous woman dancer performed dances using veils dipped in fluorescent salts containing radium. . . . Radium roulette was popular at New York casinos, 'featuring a roulette wheel . . . washed with a radium solution, such that it glowed brightly in the darkness.' A patent was issued for a process for making women's gowns luminous with radium, and Broadway producer Florenz Ziegfeld snapped up the rights for his stage extravaganzas."

The Radium Chemical Company of New York City ran ads in professional journals in which it listed radium's effects internally as improving blood pressure, increasing urine secretion, helping with arthritis pain, and producing "a general stimulation." Patients received radium injections to treat high blood pressure, menstrual cramps, depression, and an ailment that doctors labeled "debutante's fatigue." By the 1930s, hundreds of radium-based products, including eyewashes, suppositories, and even candies, were available to the public. A Harvard graduate and con artist named William John Aloysius Bailey claimed his patent medicine, Radithor, which contained two microcuries—a millionth of a curie, now the standard measure of radioactivity—of radium, was a powerful aphrodisiac that would make women "look much fresher and become more slender." Men who drank Bailey's elixir—which at a dollar per bottle was popular only with the ultrarich—would experience "a sexual rejuvenescence."

During these days, radium was completely unregulated by the federal government because a law passed in 1906 deemed it to be a naturally occurring element, not an artificial drug. But it was subsequently determined that virtually all medical applications of radium amounted to pure quackery. This was temporarily obscured by the fact that the body defends itself against radiation by producing extra red blood cells, which can give the appearance of vibrancy and good health—until radiation-damaged cells outnumber and overpower healthy ones. Radium's only truly effective medical use is in cancer therapy. Beginning early in the century, a technique was developed in which tiny needles filled with radium were used to kill cancer cells. (The theory holds that controlling the dose of radiation allows for the selective irradiation and elimination of cancerous cells.) Even this procedure entailed risks, though, and during the past quarter century radium has mostly been dropped in favor of safer techniques involving radioactive isotopes of cobalt and cesium.

X rays had demonstrated that external sources of radiation could lead to grave health problems. In the mid-1920s, radium provided the first signs that ingesting radioactive materials could be equally deadly. At that time, American newspapers were filled with stories about a group of former workers, all women, at a U.S. Radium plant in East Orange, New Jersey, who were dying from radioactive poisoning. The cause was Undark, a radio-luminescent paint used to make clock dials glow in the dark.

Undark contained thirty thousand parts of zinc sulfide to one part of radium—a mixture that had been invented a decade earlier by Dr. Sabin von Sochocky, an Austrian physicist. He predicted that the homes of the future would be painted with radium, which would bathe rooms naturally in a "soft moonlight." Businessmen quickly recognized Undark's commercial potential. Within a few years, American soldiers in World War I went into battle with glow-in-the-dark watches. Undark was also used to coat instrument panels of military aircraft to facilitate night flying.

By the early 1920s, there were fifty radium-paint studios in the United States employing several thousand workers. As recounted by Catherine Caufield in *Multiple Exposures: Chronicles of the Radiation Age,* most of the workers were young and lured by the high salary of eighteen dollars per week. They produced radium-lit fish bait, theater-seat numbers, doll eyes, and stick-on locator buttons that could be applied to bedposts, slippers, and water glasses left on bedside tables.

Beginning in 1924, dozens of radium painters began to fall

mysteriously ill and die. Doctors initially diagnosed them as
having perished from anemia, but an insurance-company
statistician who gathered data on the workers at the East Orange
plant determined that the women had died from swallowing
radium paint. The primary cause was a practice called lip
pointing, whereby workers used their mouths to make a fine tip
on their brushes. Many had absorbed more radium from
painting their fingernails and teeth to glow in the dark to
surprise friends. Radium was at this time one of the most
expensive substances on the planet, costing about three million
dollars per ounce; about seventy cents' worth was enough to
provide a fatal dose. Workers absorbed so much radium that
when they exhaled onto a zinc-sulfide screen it glowed. More
remarkable yet, some ingested so much radium that their grave
sites still caused Geiger-counter needles to jump sixty years
later. Edwin Lehman, U.S. Radium's chief chemist, also died of
cancer caused by inhaling infinitesimal amounts of radium-
laced dust. His bones were so radioactive that, wrote Caufield,
"left on an unexposed photographic plate, they photographed
themselves."

The case of the radium-dial women led to a lawsuit in 1928.
Many of the victims were so depleted by cancer that they had to
be carried to the witness stand. One couldn't raise her hand to
take the oath. During the court proceedings, U.S. Radium argued
that Undark had nothing to do with the workers' deaths.
Company lawyers insisted that radium was harmful only in
higher doses; the smaller amounts that the women had ingested
would have had "a beneficial and not a baneful effect." The
company suggested that the women's problems were

psychological, saying that radium, "because of the mystery which surrounds much of its actions, is a topic which stimulates the imagination, and to our mind, it is to this and not to actual fact that many of the reports of the luminous paint's effects in our plant may be attributed." The company agreed to an out-of-court settlement, which provided each victim with a ten-thousand-dollar cash payment, a six-thousand-dollar annual pension, and reimbursement of medical costs. The women's husbands received a small sum as well for the "loss of their wives' services." Still, the company denied all responsibility for the illnesses and said it had acted purely out of "humanitarian considerations."

Six months after the case was settled, von Sochocky, the paint's inventor, died. "His bone marrow, destroyed by radiation, had stopped producing blood cells," Caufield wrote. "The radium inside him had also eaten away at his hands, his mouth and his jaw."

Following the lawsuit, working conditions were improved at radium-paint studios. Workers were required to wear gloves, the paint was mixed in an enclosed container to avoid release of fumes, and the radium itself was stored in lead containers. But there was no way to keep radium particles completely out of the work environment. The last radium-painted watches were made in 1968, replaced eventually by phosphorus, which begins to fade after five years.

Meanwhile, other commercial products containing radium were also coming under attack. In 1932, Eben Byers, a tycoon and socialite who had been the poster child for Radithor, consuming several bottles a day and even offering it to his racehorses, died of radium poisoning. Two years earlier, Byers had stopped drinking the beverage after his teeth started falling out. Nevertheless, by the following year his upper jaw and most of his lower jaw had been eaten away. (Bailey, Radithor's creator, died of cancer in 1949.)

In 1938, the German chemist and physicist Otto Hahn was firing neutrons at uranium nuclei. His experiments ultimately succeeded in splitting the uranium atom, thereby demonstrating the potential of chain reactions—and rendering atomic power (and the bomb) a real possibility rather than a pipe dream. Scientific euphoria again bubbled over; it was as if the deaths and disease caused by radium had never occurred.

In 1940, *Collier's* ran a lengthy article by R. M. Langer, a physicist at the California Institute of Technology, which breathlessly foretold of the uranium-powered future. Energy would be so cheap "it isn't even worth making a charge for it," he predicted. Meanwhile, roads and highways would disappear, as Americans would travel in gigantic uranium-powered vehicles with enormous, soft tires that would spare the countryside. For long-distance travel, families would climb aboard their personal atomic airplanes, which would soar fifty miles above the earth and reach speeds of several thousand miles per hour.

This vision of a brave new world was not an unrealizable utopia, said Langer, but "a statement of facts that will profoundly change for the better the daily lives of you and yours. . . . The foundations of the happy era have already been laid." For one and all, the nuclear future promised unparalleled richness and opportunities. Privilege and class distinctions and other sources of social uneasiness and bitterness will become relics because things that make up the good life will be so abundant and inexpensive. War itself will become obsolete because of the disappearance of those economic stresses that immemorially have caused it. Industrious, powerful nations and clever, aggressive races can win at peace far more than could ever be won at war.

But the atom's first contribution to military history was not to eradicate war but to serve as the foundation for the first weapon of mass destruction. In 1942, with Allied forces reeling from a string of defeats before Hitler's troops, the American government mobilized the greatest collection of physicists, chemists, and engineers ever assembled, to design and build an atomic weapon. Led by J. Robert Oppenheimer, a professor of physics at the University of California and the California Institute of Technology, the Manhattan Project's research team numbered in the thousands; total personnel at the peak topped 125,000.

The central question to be addressed by the scientists was whether uranium was capable of a sustained chain reaction. This was proved by Enrico Fermi, the Italian scientist who in 1938 had traveled from Rome to Stockholm to receive the Nobel Prize for his experiments with radioactivity. Instead of returning to Mussolini's Italy, Fermi took a boat to the United States. Four years later, he led a team of scientists that achieved the first sustainable nuclear chain reaction with uranium, in a makeshift laboratory set up under the stands of a football field at the University of Chicago. (Even then, the scientists were fearful of what they had unleashed. One of the men on hand, Leo Szilard, said, "This day would go down as a black day in the history of mankind."). The chain reaction took place in a small reactor, which Fermi called an atomic pile. One of the control rods made of cadmium—which blocks neutrons, which are used to split the uranium atom and initiate the chain reaction—was attached to a rope over a pulley and suspended above the reactor. Should something have gone wrong, a scientist named Norman Hilberry was to cut the rope with an ax, thereby dropping the cadmium rod into the reactor and, it was hoped, halting the chain reaction before a meltdown occurred. Hilberry's job title was Safety Control Rod Ax Man; hence, ever since then, an emergency nuclear-plant shutdown has been called a scram.

Scientists still had to grapple with a huge problem—namely, securing enough pure uranium to make a bomb. Uranium itself is quite common in nature—more common than silver, for example—but only a variant known as uranium-235, which accounts for just 0.7 percent of the overall content, is capable of undergoing a chain reaction.

Separating U-235 from uranium ore is an extremely difficult task. Manhattan Project chemists developed two major separation processes, one using electromagnetic force and the other gaseous diffusion. With the former method, which is the most widely used today, a form of uranium gas is fed through centrifuges. As the centrifuges spin, the heavier U-238 gravitates toward the rotating outer wall of the centrifuge, while the U-235 remains closer to the axis. In a gaseous-diffusion operation, uranium gas is pumped through a series of cylinders with porous walls. The lighter U-235 passes through at slightly greater rates.

Initially, these methods produced tiny specks of U-235 that were barely visible to the naked eye. It took two years before Oppenheimer's scientists had isolated enough of the shiny, whitish metal for a bomb core.

Obtaining sufficient fissionable material is a complex task, but building an atomic bomb is not. Step one, according to John Pike—director of GlobalSecurity.org, a Washington area national security watchdog—is to assemble a bomb core of uranium (or plutonium, the other fissionable material capable of instigating an atomic explosion). The core is surrounded by a sphere of high explosives, which, when detonated, trigger a shock wave that compresses the fissile material into a critical mass that initiates a self-sustaining chain reaction. In other words, as the uranium atoms fission, or split in half, they generate neutrons that cause more and more uranium atoms to fission. The intense heat and energy thereby generated bursts the core and produces an atomic blast. The entire process takes place in roughly one-millionth of a second.

The world's first atomic blast took place during the predawn hours of July 16, 1945, at a test site in the basin of New Mexico's Jemez Mountains. The result stunned even the Manhattan Project scientists on hand, who watched—from bunkers ten thousand yards away—a mushroom cloud shoot thirty thousand feet into the air. It was reported that the bomb's blast emitted such intense light that some people living fifty miles away were certain that the sun had risen twice that day. Even more astonishing is that the light penetrated the consciousness of a blind girl who lived more than one hundred miles away. Reactions among the people who created the atomic bomb were a conflictive mix of euphoria and dread. "I am become Death," said Oppenheimer, quoting from the Bhagavad Gita, "the destroyer of worlds." Test director Ken Bainbridge remarked, "Now we are all sons of bitches."

Less than a month later, the U.S. warplane *Enola Gay* dropped a four-and-a-half-ton uranium bomb nicknamed Little Boy on Hiroshima, Japan, at 8:15 A.M., August 6, 1945. One can read and reread accounts of Hiroshima's destruction and never become inured to the horror that befell the city. In the flash of an instant, sixty-six thousand people were killed, and sixty-nine thousand people were injured, many mortally. The temperature at ground zero immediately reached 5,400 degrees Fahrenheit, and everyone within a half-mile radius was burned alive. "Birds ignited in midair," Richard Rhodes wrote. "Mosquitoes and flies, squirrels, family pets crackled and were gone. The fireball flashed an enormous photograph of the city at the instant of its immolation fixed on the mineral, vegetable and animal surfaces of the city."

Many people thought that the world had come to end—that the destruction that surrounded them had nothing to do with the war but that the earth itself had collapsed. Michihiko Hachiya, the director of a local hospital, kept a diary of the day's events. One witness who came to the hospital told him:

I saw nothing that wasn't burned to a crisp. Streetcars were standing at Kawaya-cho and Kamiya-cho and inside were dozens of bodies, blackened beyond recognition. I saw fire reservoirs filled to the brim with dead people who looked as though they had been boiled alive. In one reservoir I saw a man, horribly burned, crouching beside another man who was dead. He was drinking blood-stained water out of the reservoir. . . . In one reservoir there were so many dead people there wasn't enough room for them to fall over. They must have died sitting in the water.

Some were chastened by the effects of the new weapon. "Using atomic bombs against Japan is one of the greatest blunders of history . . . from the point of view of our moral position," Leo Szilard wrote a friend the day after Hiroshima was destroyed. "It is very difficult to see what wise course of action is possible from here on." A month earlier, Szilard had circulated a petition to Manhattan Project scientists, opposing use of the bomb. Among those who refused to sign was Edward Teller, who later built the first hydrogen bomb. In replying to Szilard, he wrote:

First of all let me say that I have no hope of clearing my conscience. The things we are working on are so terrible that no amount of protesting or fiddling with politics will save our souls. . . . But I am not really convinced of your objections. I do not feel that there is any chance to outlaw any one weapon. . . . Our only hope is getting the facts of our results before the people. This might help convince everybody that the next war would be fatal. For this purpose actual combat-use might even be the best thing.

For the most part, the destruction of Hiroshima did little to dampen the can-do spirit of the nuclear crowd. Even as the city lay in flames and ruins, Secretary of War Henry Stimson announced that nuclear fission would ultimately enrich "our civilization." "It appears inevitable," he said, "that many useful contributions to the well-being of mankind will ultimately flow from these discoveries when the world situation makes it possible for science and industry to concentrate on these aspects." It's likely that a sense of collective guilt also lay behind this post-Hiroshima spin. David Lilienthal, the first chairman of the Atomic Energy Commission, wrote years later that scientists and government officials expressed the grim determination that "the discovery that had produced so terrible a weapon simply had to have an important use."

Soon after news the bombing of Japan, John O'Neill, science editor of the New York *Herald Tribune*, rushed *Almighty Atom: The Real Story of Atomic Energy* into print. It bubbled with the same utopian spirit that marked so much of the prophetic writing of the early atomic age. "The blast of the first atomic bomb ushered in a new era of civilization, a new power age," said the book. "When you realize that man may now obtain sufficient energy from a small amount of Uranium 235 to run an automobile for 50 years without stopping, to keep an airplane in the air indefinitely, to drive the Queen Mary on countless transoceanic trips, to heat homes for years, it does not seem an overstatement to call the release of atomic energy the biggest news in history." According to O'Neill, the oil and coal industries would soon go the way of the dinosaur and "lubritoriums" would replace gas stations. Once scientists devised a safe radioactive-fuel tank— a minor hurdle that O'Neill predicted would be quickly overcome—atomic-powered planes would fill the skies.

What is striking about O'Neill's and other similar love poems to atomic power is that as yet there existed no industrial technology to harness that force. Yet the awesome power and potential of uranium seduced normally reserved scientists and intellectuals, who longed to develop this force of energy.

Beyond the need to overcome technological hurdles, atomic advocates were faced with another significant shortcoming. At the time, there were only three sources of uranium, all foreign: Czechoslovakia, Canada, and the Congo, the last being the chief supplier. Meanwhile, the government needed sufficient stocks of uranium to keep the A-bomb (and emerging A-power) industry humming. It launched a nationwide publicity campaign urging citizens to head for the hills and find domestic sources of uranium.

The poster family of uranium mining was the Steens— Charlie, Minnie Lee, and their four sons—who became cold-war heroes known across the land. A geologist from Texas, Charlie determined that an area of Utah where Marie Curie had once hunted uranium for her experiments looked highly promising. The Steens had no luck initially and lived out of a tar-paper shack, but after four years, in 1952, they hit the jackpot in a spot called Yellow Cat Wash, which sat near the town of Cisco and its thirty inhabitants. Their find triggered a rush to the area. Yellow Cat Wash ultimately produced more than one hundred million dollars' worth of uranium—as well as an epic feud along the lines of *The Treasure of the Sierra Madre*, which ended up in court and tore apart the Steen family. It also produced a mammoth environmental disaster, as millions of tons of radioactive tailings and waste piled up in the region over the decades. As of 2003, the Department of Energy was still determining when and how to haul the waste away from a dump site near the banks of the Colorado River.

The government didn't stop buying uranium from prospectors until the mid-1960s, by which time it had sufficient stockpiles. By then, some fifteen thousand men had worked in the region, creating boomtowns in the desert and turning Salt Lake City, a hub of the uranium trade, into a town of speculators and grifters. Today, the region is littered with ghost towns of the old glory days. At least five hundred prospectors subsequently died of lung cancer.

The government also desired stockpiles of plutonium, a highly fissionable element discovered (and created) by Manhattan Project scientists and used for fuel in the atomic bomb that destroyed Nagasaki three days after Hiroshima was leveled. A dull, silver-colored metal more than twice as dense as lead, plutonium is produced by bombarding nonfissionable uranium-238 with neutrons. It is one of the most deadly substances known to man. A particle the size of a speck of dust can cause lung cancer when inhaled. Plutonium is so highly reactive that small shavings of it can self-ignite, and it must be stored in small pieces to prevent spontaneous chain reaction.

Two scientists were exposed to critical levels of plutonium while working on the Manhattan Project; both died shortly thereafter. "If there ever was an element that deserved a name associated with hell, it is plutonium," Robert Wilson, a scientist who helped oversee the government's development of nuclear energy, once said.

Plutonium became the fissionable material of choice for America's nuclear weapons designers because far less of it than uranium is needed to produce a chain reaction. Whereas a grapefruit-size ball of uranium weighing about 123 pounds (56 kilos) is needed for the core of an atomic bomb, a 24-pound (11-kilo) golf ball of plutonium will serve the same purpose.

In 1952, the government founded the Rocky Flats plant in Colorado, where plutonium was produced for America's nuclear-weapons program. The total number of employee casualties is not known, but it is believed that hundreds of workers there went to early graves. Among them was James Downing, who made plutonium buttons roughly twice the size of a hockey puck that were parts of the triggers for atomic bombs. He sustained first-degree burns during a machine fire that led to the emission of plutonium and later died of esophageal cancer. Donald Gabel operated a furnace that melted plutonium—he got paid an extra fifteen cents per hour for hazardous duty—and was frequently exposed to air that leaked from it. He was dead from a brain tumor at the age of thirty-one.

The government admitted only in 1999—ten years after the plant was shut down—that workers at Rocky Flats (and other plants involved in bomb making) had died or become ill due to radiation exposure. As many as six thousand employees or their

descendants were to be compensated with payments of
$150,000. Meanwhile, the government allocated seven billion
dollars for a cleanup of the Rocky Flats site that is expected to be
completed in 2006.

When it was learned in 1953 that the Soviet Union had
developed the H-bomb, nuclear madness reached a fever pitch
and the arms race began in earnest. That year, *The New York
Times* turned the government's nuclear-testing program into a
spectator sport when it dubbed bomb watching an "honorable
pastime." Upcoming nuclear tests were announced by the press,
and thousands of residents of western states pulled out their
lawn chairs and sipped iced tea as they observed mushroom
clouds rising over the desert from almost one hundred A-bombs
that were detonated in the atmosphere during the next decade.
Oppenheimer by now had become an opponent of atomic
warfare; after he compared the United States and the Soviet
Union to "two scorpions in a bottle," he was denounced as a
spineless commie sympathizer.

Before long, though, public fears about a possible nuclear
war with the Soviet Union led to a vibrant political movement
calling for abolition of the bomb. The antinuclear cause was
most clearly espoused by Allen Ginsberg, who wrote in his poem
"America," "Go fuck yourself with your atom bomb," and later
by Stanley Kubrick, in his film *Dr. Strangelove; or, How I Learned
to Stop Worrying and Love the Bomb,* which ends with Slim Pickens
bronc-riding the nuclear weapon that triggers Armageddon.

The government insisted there was little to fear and assured
Americans that just about everyone would survive a nuclear
attack anyway, as long as they took appropriate precautions. The

Office of Civil Defense (OCD), a subbranch of the U.S. government, taught Americans the importance of bomb shelters and had as its mantra the phrase "Duck and cover." In a video the OCD produced for schoolchildren, a young boy named Tony is seen pedaling his bike down a small-town street, heading for a Cub Scouts meeting. "Tony knows the bomb can explode any time of the year, day or night," says the narrator. "He is ready for it." Suddenly, the screen turns white, and Tony jumps from his bike and throws himself to the curb. "Duck and cover!" the narrator says cheerfully. "Attaboy, Tony!"

The commercial nuclear age opened in 1957, when the first atomic-energy plant opened in Shippingport, Pennsylvania, with backing from the government's Atoms for Peace program. Even before the plant opened, nuclear-power advocates were confidently predicting that household appliances wouldn't simply run on atomic energy but in some cases would have their own internal nuclear generators. "Nuclear powered vacuum cleaners will probably be a reality in ten years," Alexander Lewyt, president of Lewyt Vacuum Cleaner, wagered around this time.

In 1959 and 1960, the Rockefeller Panel reports were published, having been overseen by Dean Rusk, Henry Kissinger, and other members of the political elite whom the historian David Halberstam would subsequently dub "the best and the brightest" in his book of that title. The Rockefeller experts were concerned about sustaining high rates of economic growth, and they believed atomic power would make that possible. "Even now we can discern the outlines of a future in which, through the use of the split atom, our resources of both power and raw materials will be limitless," they wrote. "In the 20th Century, the unprecedented acceleration of scientific advance promises that we are on the threshold of a new age of science. . . . Already, the proven resources of uranium and thorium, in terms of energy equivalent, are at least 1,000 times the world resources of coal, gas and oil."

Three years later, the National Academy of Sciences completed a report for President Kennedy that endorsed the conclusions of the Rockefeller panelists. The National Academy said that nuclear power would allow the United States to shift from a philosophy of conserving scarce resources to a policy described as "the wise management of plenty." As Secretary of Interior Stewart Udall later wrote of the study, "It cemented the census about technology and implied that, if we ran out of petroleum or iron ore—or any other mineral—technology would soon come forth with a better, cheaper substitute."

Other than the cartoon illustrations, there's little that distinguishes the reports of the Rockefeller panel and the National Academy from *The Golden Book of Chemistry Experiments* so treasured by David Hahn. And while it's easy to understand an adolescent boy's boundless enthusiasm for atomic power, the starry-eyed optimism of all those sober experts who preceded him is harder to explain.

CHAPTER 3

Burning Down the House: Basement
Explosions and Other Early Developments

It is possible that you may be permitted to work at the
kitchen table when this is not in use. But it is far better if
you have a place where you will not be disturbed and where
you can store your equipment—a corner in your room, or
in the basement or the garage.

—THE GOLDEN BOOK OF CHEMISTRY
EXPERIMENTS, 1960

he illusionist Penn Jillette (of the team Penn and Teller)
once said that he mastered the art of juggling at a very
young age—adding that this achievement was another way of
saying, "I have a terrible social life. I'm not normal. I spend all

my time practicing." By the time he began his freshman year at Chippewa Valley High School in 1990, just a few months shy of his fourteenth birthday, David, who spent all of his time practicing science, could say the same.

Chippewa Valley High was located a few blocks from David's father's house in Clinton Township and had a student body of several thousand. In downtown Detroit, many high schools had installed metal detectors at the front door in an effort to keep kids from packing heat along with their lunches. Chippewa Valley had its share of drugs and school-yard fights, but compared with inner-city schools it was a placid, well-endowed institution. The school boasted classrooms stocked with the latest equipment, ample facilities, and a spacious football field for its team, the Big Reds (motto: "We are warriors, but we are not ordinary warriors"). The latter held absolutely no allure for David, who didn't fit comfortably into any of the high school cliques, least of all the jocks.

David was never an enthusiastic student, and despite his obvious talent for science—indeed, precisely because of that talent—he quickly fell behind in his studies. He frequently sat in the back of the classroom and feigned interest in what his teachers were saying, while concentrating on reading material he kept hidden behind his class books. For most adolescent boys, that reading material would have been the latest edition of *Playboy*, a history of baseball, or a copy of *Rolling Stone*. David was secretly absorbed in a chemistry book or a magazine account of the latest developments in the nuclear-power field.

The influence of this reading, particularly the supreme ability it ascribed to scientists to vanquish the nation's economic, social, and ecological woes, seeped into David's schoolwork. "Ever since the beginning of time man has dumped his waste into our environment," he wrote in one of his research papers. "Today all of this waste is building up to large amounts." Luckily, David foresaw a simple solution: biodegradable chemicals that would magically transform industrial waste "into harmless compounds." Despite all his research, it never occurred to David that chemicals themselves could pose a menace or that a primary cause of the "waste crisis" was the very science he invoked.

The one outlet for David's talents came, of course, in science classes. He formed a good relationship with some of the Chippewa Valley science faculty, especially Ken Gherardini, his physics teacher, whom he looked upon as a mentor. Yet even Gherardini, though fond of David, thought his student was pretty flaky and remembered him as having a very brief attention span, especially when class discussion veered away from his pet interests.

David received better, if not exceptional, grades in science than in other areas. What made him stand out, though, was his eccentricity and cockiness. One day, Ken Hahn received a phone call from David's biology teacher, who told him that his son repeatedly interrupted her class lectures. But the teacher was more amused than angry: She had become so exasperated by David's interjections during her talk about acids and bases that she told him to come to the front of the room and finish the lecture for her. David did, ably; he enjoyed it so much that he asked if he could give another lecture the following week.

David subsequently once convinced his biology teacher to let him demonstrate a process for acid-base neutralization that called for mixing equal quantities of hydrochloric acid and sodium hydroxide. If done properly, this produces an exothermic reaction, one that produces a great deal of heat (as opposed to an endothermic reaction, which absorbs heat). David strode to a lab table, mixed the chemicals in a beaker, and waited for nature to take its course. Unfortunately, he misjudged the proportions, and the reaction was too strong. The solution began boiling over, and a thick cloud of smoke filled the classroom. The reaction did simmer down on its own, but not before splashing over the side of the beaker and burning the lab table. Afterward, the kids treated the whole thing as a joke—just another example of David's increasingly weird behavior—but during the experiment, when it looked like the chemicals might erupt, a few students had bolted out the door.

In his spare time, David lapped up the fundamentals of atomic history, his new primary passion. Elements, he discovered, are distinguished by the number of protons in their nuclei. Hydrogen, the lightest element, has one proton; hence, its atomic number is 1 and it is first on the periodic table of elements. Helium, number 2 on the table, has two protons in each nucleus. Most nuclei also contain neutrons, though the number varies. The combined number of protons and neutrons (called nucleons) in a nucleus is the atomic weight or mass.

Elements generally occur in different forms, or isotopes, which are distinguished by the number of neutrons in their nuclei. For example, there are three isotopes of hydrogen: H-1,

H-2, and H-3. The first has one proton and no neutrons; the second, called deuterium, has one proton and one neutron; and the third, tritium, has one proton and two neutrons.

Most elements have at least two naturally occurring, stable isotopes. But isotopes of heavier elements (those with more protons) are often unstable. Called radioisotopes, they give off energy—alpha, beta, or gamma—until they revert to a more stable form, a process called radioactive decay, which produces radioactivity. Uranium has ninety-two protons in its nucleus—hence, it is number 92 on the periodic table—and its isotopes range from U-232 to U-238.

David learned that all radioactive elements and isotopes have a half-life, a term that refers to the amount of time required for the intensity of their radiation to decay by half. Uranium has a half-life of 4.5 billion years, while radium's is 1,620 years. Se-82, a man-made isotope of selenium, decays so slowly that scientists estimate its half-life at ten billion times the age of the universe. Other radioactive isotopes have a half-life of less than one second. An element's half-life reflects the intensity of its radioactive emissions, but there is no direct correlation between half-life and lethality to people. For example, plutonium has a half-life of 24,000 years, while thorium has a half-life of 14 billion years; both are highly dangerous to humans, but the former is considered to be far deadlier than the latter.

Meanwhile, David continued his research into the Curies. He read everything he could find about the couple. He was moved just as much by their hard work with crude tools as he was by their genius. David would soon employ methods just as crude in his efforts to emulate his heroes.

David also gobbled up information about the other nuclear pioneers. Among those he especially revered was his namesake, Otto Hahn, the man who first split the uranium atom; Sir James Chadwick, who discovered the neutron; and Frédéric and Irène Joliot-Curie (the latter the daughter of Marie and Pierre), who received the 1935 Nobel Prize in chemistry for producing the first artificial radioisotope by bombarding nonradioactive elements with alpha particles. It was later discovered that neutrons would be more effective than alpha particles in creating radioisotopes, but the Joliot-Curies' discovery is still considered to be one of the most important of the twentieth century. (The couple showed their handiwork to Marie Curie, who was then dying. "I will never forget the expression of intense joy which came over her face when Irène and I showed her the first artificial radioactive element in a little glass tube," Frédéric wrote. "This was doubtless the last great satisfaction of her life.")

With heroes like these, it's easy to understand David's unbridled enthusiasm for the atomic fellowship. The trouble was that the radioactive age hadn't been nearly as carefree and painless as David's romanticized vision. Alpha radiation does not pose a serious external threat to people, as it travels only a short distance and does not easily penetrate objects or the human body. If alpha-emitting substances are inhaled or ingested, though, they can be deadly. Beta particles can travel up to about six feet and pass through an inch of human tissue and therefore represent an external hazard, especially for skin burns. Gamma rays pose the greatest external threat, because they are highly penetrating and can irradiate the entire body. Gamma emitters are equally dangerous if inhaled or ingested.

Many of the nuclear pioneers died as result of their labors. Marie Curie's exposure to radium led to her death in 1934 from leukemia. By then, she was exhausted by radiation sickness, nearly blind, and suffering constantly from dizziness and fever. Pierre had died twenty-eight years earlier, run over by a horse-drawn cart near the Pont Neuf in Paris. Otherwise, he would surely have suffered an end similar to his wife's; at the time of his death, his hands were already badly damaged by radiation burns, and he showed other signs of radiation poisoning. The Curies' scientific papers absorbed and emitted so much radiation that even today visitors to the Bibliothèque Nationale who wish to examine the couple's notebooks must sign a waiver assuming all health risks.

If he'd been interested, David would have found it a simple matter to learn about the cons as well as the pros of the atomic age. But as with newfound converts to any cause, David wanted his ideas reinforced, not challenged. Certainly, the Curies and some of the other nuclear pioneers had suffered as a result of their labors, but that hadn't kept them out of the laboratory. For David, as for his heroes, the thrill of discovery made worthwhile any risks.

David pored over his nuclear research whenever he had a free moment, be it in the hallway walking between classes or at a cafeteria lunch table. A few Chippewa Valley students were curious about David's scientific interests, but the vast majority thought he was simply a geek. Some of his schoolmates dubbed him Dork Boy, and David occasionally had fistfights with his tormentors. After one altercation, he asked Ken to buy him a weight set. From that day on, he worked out regularly and before long gained enough bulk and confidence that bullies left him alone.

David had a small group of friends at Chippewa Valley. Not surprisingly, they tended to be science fanatics and were frequently, though not always, loners and outcasts as well. There was Spencer Hawkyard, a technology geek and computer wizard; Jim Miller, a skinny, nervous kid who lived near David's house in Clinton Township and shared his passion for all things nuclear; Andy and Jeffrey Hungerford, brothers two years apart who were highly knowledgeable about chemistry; and a few kids from his local scout troop.

At school, David displayed little interest in student clubs, team sports, and other after-school activities. Apart from his class portrait, he hardly appears in the yearbooks. After the final school bell rang, David usually hurried to the library, to his home lab at Ken and Kathy's, or to the potting shed in Commerce Township.

The shed was an enormous source of pride for David. In his mind, it was the mirror image of the one where Marie and Pierre Curie had discovered radium. Truth be told, David's lab was indeed similar to theirs. Both were rickety structures with dirt floors that offered little protection from the elements (whether climatic or radioactive). Soon after the Curies discovered radium, German chemist Wilhelm Ostwald traveled to Paris to see their laboratory. The couple was out when he arrived, but

Ostwald was so excited to view their scientific shrine that he prevailed upon an assistant to give him a tour. The chemist was astounded. "It was a cross between a stable and a potato shed," he later said. "If I had not seen the worktable and items of chemical apparatus, I would have thought that I was [the victim of] a practical joke."

Like the Curies' shed, David's was a sauna during summer and an icebox during winter. David was determined to keep his shed in operation year-round, no matter how much rain, sleet, and snow he had to confront during Michigan's terrible winters. He'd bundle himself up in a heavy coat, hat, gloves, and scarf and work until his teeth were chattering. One winter, though, the cold was so unrelenting that the shed became intolerable. As a final, desperate measure, David set up a charcoal grill inside the shed, but with no ventilation system the smoke—and carbon dioxide—drove him out. Even David had to concede defeat and shut down until the spring thaw.

David's seclusion in the shed was further spurred by the isolation and alienation that marked his family life. Michael and Patty's relationship was turning increasingly sour, and the house in Commerce Township was frequently rocked by disputes between the two. Fueled by too much booze, their fights sometimes turned violent enough that David had to call the police. One especially ugly confrontation ended with Patty taken to the police station in handcuffs, though no charges were pressed. On several occasions, David packed up his bags as his mom and Michael raged. Knowing that the cops would invariably turn up, he'd call his father and wait on the front porch until Ken drove out and picked him up.

In Clinton Township, Ken's habit of working into the night made dinnertime a lonely affair. Kathy would put plates of food on the kitchen table, but she, Kristina, and David often scattered with their meals. David carried his plate to his room and read while he ate; Kathy and Kristina watched TV in the living room.

Concerned that David was becoming overly consumed with his experimenting, Ken pushed him to find an after-school and weekend job. He thought work would distract David and keep him too busy for science. David flipped burgers at a McDonald's, washed dishes for a hot-dog joint, loaded furniture on trucks for La-Z-Boy, and bagged groceries at a Kroger's supermarket—but work was merely a means of financing his scientific endeavors. He'd often call in sick to work or just not show up, then sneak off with a backpack full of chemicals, test tubes, and beakers to the tree-house fort that he and his friends had built years earlier. There, he could mix fireworks and carry out other simple experiments without fear of being discovered by Kathy, who was becoming ever more watchful of his activities.

When he did punch the time clock, David's attention frequently drifted from his duties. He'd use his break time at Kroger's to analyze the content labels of hundreds of products, from dishwashing detergents to floor cleaners to cold medicines. When he found an alluring chemical in a product, he'd try to isolate it or produce it in his home lab.

One day at Kroger's, another employee was wheeling a fully loaded cart through an aisle when eight large containers of ammonia toppled from it and shattered on the floor. The fumes spread rapidly throughout the store, threatening to overcome shoppers. Inhaling too much ammonia causes blisters to form in the throat and nostril hair to fall out; prolonged exposure can even be fatal. Customers headed toward the exits, but a handful struggled for breath and rubbed their streaming eyes. One older man looked to be in serious trouble.

In his head, David had carefully charted Kroger's for its chemical components. Knowing that ammonia is a base that can be neutralized with an acid, he rushed to an aisle that carried The Works, a toilet-bowl cleaner that contains a high level of hydrochloric acid. Grabbing a bottle, he dashed back across the store and poured the cleaner on the ammonia. A huge white cloud immediately formed and spread across the supermarket— but the smell of ammonia dissipated almost instantly.

David thought he would be acclaimed the hero of the day. His boss, however, thinking that the white cloud was a toxic vapor, was irate. He readied an evacuation of the store and fired David on the spot. Fortunately, another store employee called poison control and learned that David had done precisely the right thing. His contrite boss apologized and offered him his job back, which David graciously accepted. (David described that day's events many years later when he took me to that Kroger's during a tour of his former haunts. As he walked through the aisles, David grabbed various products off the shelf and analyzed their ingredients. "Six dollars and eighty-nine cents," he sniffed disgustedly upon examining the contents of Just For Men hair dye. "I could make that for twenty cents.")

David's social life was limited, but he did maintain a healthy attraction to girls. (This was another manner in which he differed from Rhodes's composite scientist, who displayed little interest in the opposite sex.) And though he may have been a

science nerd, he was a handsome one, not the stereotypical
teenage brain with thick glasses and braces. With age and weight
lifting, the angelic-looking boy of a few years earlier had become
a sturdy adolescent. By the time he was a junior in high school,
David was a well-trimmed five foot nine, wore his short blond
hair combed back and parted neatly on the right, and had a big,
irresistible smile. His wardrobe had also changed, from jeans
and sneakers to slacks and dress shirts.

After a few brief flirtations, David began dating Heather
Beaudette, an intelligent, pretty, brown-haired girl who loved to
paint her long fingernails startling colors. Heather hit David like
a thunderbolt, the same effect Patty had had on David's father
when he spied her for the first time on Woodward Avenue. The
very night that David met Heather, a chance encounter at the
house of a mutual friend, he asked if he could kiss her. She
agreed—it was the first real kiss for either of them—and the two
were soon inseparable.

David's family background hadn't left him particularly well
prepared for romance, and Heather was the first person to whom
he showed real affection. Yet he was a sweet and caring
boyfriend. Heather returned from one weeklong family trip to
find a pile of lengthy love letters from David, who clumsily but
sincerely poured out his feelings for her and for once managed
to stay off the subject of science.

As a rule, though, David's fixations made him an awkward
date, and his conversational style ran to lecturing. During the
time when food and diet obsessed him, David was downing daily
doses of his mushroom shakes and eating huge amounts of
kiwifruit, which contain L-tyrosine, an enzyme believed by some

researchers to enhance memory. At dinner with Heather and her parents, he'd spend the entire meal analyzing the chemical makeup of the prime rib or beef Stroganoffs. "I couldn't get him to shut up," Heather recalled.

Heather's mother, Donna Bunnell, thought David was peculiar but judged him, at least initially, to be responsible and certainly preferable as a boyfriend to some of Heather's more rowdy suitors. When David asked her to write a letter of recommendation for a scouting project, she happily agreed. Her letter praised him as being polite and responsible and noted that he always brought Heather home from dates at the agreed-upon time. Still, Donna didn't know what to make of David's idiosyncrasies. She'd try talking to him about current events or sports, but he had a maddening way of linking every subject to science. Donna invited David to her second wedding as Heather's date but tutored him sternly on how to behave during the reception. "He was a nice kid, good-looking, and always presentable, but we had to tell him not to talk to anybody," she said of the drills she ran him through. "He could eat and drink but, for God's sake, don't talk to the guests about the food's chemical composition."

Donna knew that David was conducting some sort of scientific "research," but she assumed it amounted to little more than making water turn different colors. One evening, though, she returned home from a computer class just as her husband, Allan, and Heather were running out the front door. David was at home alone, they told her excitedly while dashing for the car, and had suffered burns during an explosion. Donna was petrified. She imagined a natural-gas explosion had ripped through the Hahns' house, but it turned out that the detonation had resulted from another chemical experiment gone awry. Happily, David's injuries were relatively minor.

A short time later, Donna took Heather aside and told her she should seriously consider what type of future was in store for her if she stayed with David. Heather wasn't about to let her mom pick her boyfriend, but she wisely concluded that the topic of David's scientific activities should best be avoided. Whenever David tried to chat with her mom about his experiments, Heather would roll her eyes and tell him to knock it off.

For her part, Donna had concluded that the entire Hahn household was strange. She tried to reach out to Ken and Kathy by inviting them to Heather's birthday celebrations and parties at her house but never even got a reply. In fact, there was no social contact between the families at all. When Allan accompanied Heather to the Hahns' home on the day of the explosion, he returned shaking his head in disbelief. Ken had arrived there, too, but had seemed strangely blasé about the blast. He barely acknowledged Allan or Heather, and, once he had established that David wasn't seriously hurt, he walked to the kitchen to fix himself a sandwich and look through a pile of mail. Ken's manner gave the impression that household explosions were nothing out of the ordinary—which was roughly the case, though of course Allan didn't know that.

David remained seized with the idea of discovering new and improved means of tanning—the body's attempt to protect itself against solar radiation—as just a few hours in the sun would leave his fair skin a bright, painful red. He bought a set of tanning lights and mounted them on his bed, where he'd stretch out for brief spells in the hopes of slowly achieving the perfect tone. Heather remembered that David once fell asleep under the lights after applying one of his homemade celery-based lotions and emerged "red as a lobster." Did she ever volunteer to take part in David's experiments? I asked her. "No way!" she replied with a shudder. "He wanted me to, but I saw him come out burned and wrapped in gauze. I wasn't going to let him try any of that stuff on me."

David's interest in tanning led to perhaps his most bizarre set of experiments. In the course of his investigations at the local public library, he learned about melanocyte-stimulating hormone (MSH), which provokes an increase in pigmentation in humans. Unfortunately, the hormone is found only in the pituitary gland of a cow.

This piece of news would likely have dissuaded most kids, but it didn't slow down David for a second. He visited a nearby slaughterhouse that sold newly severed cow skulls to medical schools and drug manufacturers (the latter seeking, among other things, the same hormone as David was) and convinced an employee there to take him on as a client. David became a regular visitor. He paid ten dollars per cow skull—a price that included the services of a worker who scooped out the brain and cut out the pituitary gland.

The slaughterhouse kept the rest of the brain, and David departed with the pituitary glands, which he iced upon arrival at his home to keep the MSH from degrading. With the aid of *The Handbook of Vitamins, Minerals and Hormones*, which he checked out from the library, he discovered a means of isolating MSH

using potassium chloride, carboxylic acid, and other chemicals
that he obtained via mail order from stores in Texas and
California. He'd scrape off some of his own skin cells with a
sharp knife, put them under a microscope lens, and see how they
reacted to the hormone. He also applied MSH topically and
claims to have achieved good results.

His success may have been more than imaginary, as David's
tanning experiments bore some distinct similarities to ones
being developed simultaneously by "sunless tanning"
researchers in the United States and Australia, who were
conducting experiments with the hormone as well. By 2001, tests
conducted with the approval of the U.S. Food and Drug
Administration seemed to show that MSH can induce artificial
tanning. Researchers were further encouraged by signs that it
may, like Viagra, help in the treatment of male erectile
dysfunction.

Heather didn't know about David's experiments with cow
brains, but she was frightened enough by what she knew about
his backyard laboratory in Golf Manor. One day, David took her
to his mother's house to show off his experiments. His timing
couldn't have been worse. When they entered the potting shed,
they found shattered chemical containers, scorch marks on the
lab table, and general mayhem. The cause was a time-delayed
blast produced by David having mixed excessive quantities of
glycerin and potassium permanganate, a chemical combination
that he used to create smoke bombs and a highly flammable
ignition compound he used in his rocketry experiments. Adding
to the gloom was the presence of an unwitting victim. Shredded

almost beyond recognition was an unlucky raccoon that had been prowling around inside the shed when the chemicals erupted.

Oddly, Heather was untroubled by David's growing interest in radioactive materials and nuclear power—even when he mentioned casually that he was hoping to get his hands on a quantity of uranium to experiment with. "I didn't know anything about uranium, even that you weren't supposed to have it in your house," she said in recollection. "I was like, 'Whatever.' "

Meanwhile, the weird activities in the backyard occasionally attracted attention from Patty and Michael's neighbors. One saw thick smoke coming from the shed one afternoon and knocked on the door to see what was going on. David stuck his head out and said he was just doing a small welding job, though in fact he'd been testing a new smoke-bomb recipe. Another day, David decided to create chlorine gas—a version of the mass killer from World War I—following the procedure laid out in the *Golden Book*. Even David recognized that an accident in the shed could be dodgy, so he conducted the experiment on a card table set up at the side of the swimming pool. In the middle of the experiment, a neighbor mowing his lawn, undoubtedly concerned by the gas mask David was wearing, turned off his engine and walked over to the fence between the backyards.

"What's going on over there, David?" the man wanted to know. "Do your folks know what you're doing?"

"It's nothing really," David replied, ignoring the second question. "I'm just making my own oxygen."

Mollified, the man gave the thumbs-up sign and returned to his lawn work.

Though David frequently worked at the poolside table, especially during the dog days of summer when the shed was blistering hot, these two incidents were the only ones that led anyone to question him about his activities. Probably the neighbors were disarmed by his youthful innocence and beatific glow. "I'd be outside working with gunpowder or nitroglycerine, and the neighbors would smile and wave to me from their backyards," David said. "I think they thought I was just some dorky kid."

David seemed unable to forecast any bad consequences from his work. The precautions he took in the shed were laughable to nonexistent. Beyond his occasional use of a gas mask, he used no safety gear other than plastic goggles and paper-filter masks of the type used by painters or nurses. None of that would have done him much good in the event of an explosion, nor did it offer sufficient protection from chemical fumes—let alone from the radioactive elements that he was dreaming of acquiring.

The lack of concern he showed for his own personal well-being extended to that of his family as well. One day in Clinton Township, he poured the remains of a chemical experiment down the toilet and spilled some on the toilet seat. When Kathy used the toilet later that day, she suffered a painful and embarrassing burn on her behind.

Understandably, this incident prompted Kathy to scrutinize David's activities. She now began routinely searching David's room and disposing of any chemicals and equipment she found, usually hidden under the bed or buried at the back of a closet. Kathy once seized a chemical stash and locked it in her bedroom. David tried to pick the lock on the door, and when that failed he climbed up a ladder to get in the bedroom window and took it all back.

But even Kathy, who was far less indulgent of David's activities than Ken, never comprehended the scope of her stepson's activities or his ambitions. She'd see him poring over chemistry textbooks filled with equations she couldn't make sense of and listen to him talk about radioactive research but couldn't imagine he was capable of anything more than rudimentary experimentation. "I knew he was book smart," Kathy later said, "but it never occurred to me that he was capable of carrying out the [experiments] he was reading about."

David frequently woke up Ken and Kathy with late-night activities in his bedroom. First would come the sound of a running motor or generator. Then a caustic, burning smell would hit Ken in the nostrils and make him sit up in bed. He'd stomp down the hallway, throw open the door to David's room, and find his son scurrying to hide things under the bed.

Finally, Ken and Kathy banished David to the basement, thinking that he'd cause less damage down below and they'd get more sleep. That suited David perfectly. He moved his bed, dresser, bookshelves, and plants downstairs and turned the basement into a private refuge where he could read, dream, and experiment. His new lair allowed him to further distance himself from his parents and gave him greater freedom to create and destroy things, to break the rules, and to escape into something he was a success at, while sublimating a teenager's usual sense of failure, anger, and embarrassment into some really big blasts. David describes the move as a "liberating event": "My room was way too small, and the basement was a much better research environment," he said. "They left me alone down there, so I could do almost anything I wanted."

Years later, scars dating to David's reign of scientific terror were still evident in the Hahn-family basement. Numerous chemical burns had melted away sections of the linoleum floor, and the leg of a wooden table displayed charring, presumably from one of the fireball explosions that had once been as much a part of life in the household as the once-a-week family excursions to McDonald's for dinner.

David's move downstairs offered Ken and Kathy only a partial respite. The basement housed the central air-conditioning unit, and smells from David's experiments would get sucked into the intake grill and from there circulate throughout the house. Ken opened the front door after arriving from work one evening to find a smell of smoke so strong that he rushed to the phone to call the fire department. Upon reflection, he decided to check the basement first. He ran downstairs and discovered David happily and obliviously working away in a smoky haze.

Finally, even a workaholic like Ken could no longer ignore his son's strange behavior. "We tried to keep him busy with school and work so he wouldn't have time to get in trouble, but he always found a way," Ken said. "It was hard to even get my son to sit down and talk. He was always so wrapped up in his experiments. I'd say, 'Let's go out and play basketball,' but he was never interested. I asked him to try out for track or tennis or wrestling, but he never would. We had to force him to play soccer. His attention span was about ten seconds. I'd be talking to him about what time to be home at night, and he'd suddenly be going on about his experiments. I thought he might have attention deficit disorder. I went and had him tested, but the doctor said that wasn't the problem—he didn't know what the problem was."

Convinced that his experiments and increasingly erratic behavior were signs that he was making and selling drugs, Ken and Kathy began to conduct spot checks at the public library, where David told them he studied. To their surprise, David was always there as promised, usually buried behind a vast pile of chemistry books. Ken and Kathy were not mollified. Fearful that he would burn down or level their house, they prohibited David from being home alone. They forced him to accompany them on short errands; if they planned to be away longer, they'd drop off David at a friend's house or the library and pick him up when they returned. This must have been a humiliating ritual, but David showed no signs of giving up his work.

All great inventors, particularly those working with high explosives, have encountered spectacular setbacks and mishaps. The first guns often exploded in users' hands, and in 1460 an early cannon blew up and killed King James II of Scotland, who was standing nearby. Tragedy also befell Alfred Nobel of Sweden, who in the mid-nineteenth century was searching for a safe means of detonating nitroglycerin. Frequent explosions marked his years of experimenting, including one that killed his brother Emil. But Nobel finally found that a soil called claylite absorbed four times its own weight of nitroglycerin. The red powder that was created by marrying the two substances could be burned or pounded without detonating; it required a mercury blasting cap to initiate the explosion. Thus did Nobel invent dynamite—

earning enough money from its sale to help him establish the
Nobel Prizes that were bestowed on so many of the atomic
trailblazers half a century later.

David, too, had his share of catastrophes, if on a smaller
scale. One night, as Ken and Kathy were sitting in the living
room watching TV, an explosion in the basement rocked the
house. They rushed downstairs and found David lying
semiconscious on the floor, his eyebrows smoking. Unaware that
red phosphorus—which he planned to use as a detonator for a
rocket project—explodes upon heavy impact, David had been
hammering a chunk to granule size inside a plastic container.
Ken rushed David—who of course had been working without
goggles—to the hospital, where he had his eyes flushed. For
months, David had to make regular trips to an ophthalmologist
to have pieces of the plastic container plucked from his eyes, and
he had impaired vision in one eye for a full year.

This basement blast marked a turning point. Kathy flatly
forbade David from experimenting in her home and—knowing
full well that David would ignore her—gave Ken an ultimatum as
well: Either David shut down his lab, or she was moving out. That
got Ken's attention. He read the riot act to David, and the
experiments came to an abrupt halt—the ones in Clinton
Township, anyway. For David, knowing when to cut his losses,
merely shifted his base of operations to his mother's house and
the backyard potting shed.

Ken and Patty talked occasionally, but their relationship was
chilly. Perhaps that was why Ken didn't think to warn his ex-wife
or Michael that they had better keep a close eye on David. Nor
did Ken know anything about the potting shed. When Ken

dropped David off at Golf Manor, he never went inside, let alone to the backyard.

In Clinton Township, David's more reckless urges were checked at least in part by Kathy's oversight. In Golf Manor, he operated with almost complete freedom, particularly with Patty diverted by her own serious problems. Michael was briefly alarmed one day when he heard a small eruption from the backyard. He went to investigate and found David setting off explosives in the swimming pool, which from his perspective wasn't anything to get worked up about. After all, that was something he enjoyed doing as well.

Michael and Patty didn't know exactly what David was up to, but they thought it was great that he spent so much time in the potting shed. From their point of view, whatever he was doing was preferable to pot smoking and other teen diversions. "David was a good kid," Michael later said without the slightest trace of irony. "We never had to worry about him getting into trouble."

Sometimes, David tried explaining his experiments to his mom and Michael, but like everyone else they couldn't really appreciate what he was up to. "I played on like I knew what he was talking about, but what he told me went right over my head," Michael said. "I saw his notes once, and they were like Chinese to me." One thing later stuck out, though: David's work had something to do with creating energy. "He'd say, 'One of these days we're gonna run out of oil.' He wanted to do something about that."

CHAPTER 4

Radioactive Education:
Lessons from the Boy Scouts

In all ages there have been scouts, the place of the scout
being on the danger line of the army or at outposts,
protecting those of his company who confide in his care. . . .
But there have been other kinds of scouts besides war scouts
and frontier scouts. They have been the men of all ages, who
have gone out on new and strange adventures, and through
their work have benefited the people of the earth.

—THE OFFICIAL HANDBOOK FOR BOYS, 1911

While David harbored nuclear ambitions, his
experimenting to this point was still largely
unfocused. It also involved work with chemicals that are truly
dangerous only if mishandled or combined improperly. He

would soon begin to work with materials that are inherently dangerous and can cause terrible harm by passive exposure alone. Ironically, he was prodded to take that giant leap, even if inadvertently, as a result of his father's passion for the Boy Scouts.

Even as he had dropped most of his old, nonscientific pursuits, David still cherished his time with the scouts. He and other members of Troop 371 met weekly at the Veterans of Foreign Wars hall in Mount Clemens to plan and train for upcoming camping trips and outdoor activities. On the stage of the VFW, they practiced pitching a tent, learned how to start a fire in the rain (with flints, waterproof matches, and cotton balls for kindling), acted out rescue scenes in which an injured trooper was saved by his comrades, and were otherwise instructed in how to master the scout motto, "Be prepared."

One of the highlights of David's year came when he departed for an annual eight-week stay at Lost Lake Summer Camp, located in a lushly forested swath of central Michigan. Harry Bennett, during the 1930s and 1940s a top executive at Ford, originally built the camp as a retreat for himself, using equipment, money, and labor largely diverted from his employer. Local legend has it that Bennett, who was known to socialize with mobsters, used the vast estate for gambling and the production of moonshine. It was even said that foes of organized crime in Detroit often ended up at the bottom of Lost Lake in "cement shoes." Bennett, who was fired by Edsel Ford and eventually died penniless, sold the property to a steel company, which sold it in turn to the Boy Scouts' Clinton Valley Council.

David thrived at Lost Lake and ultimately became a senior patrol leader, which gave him a real sense of responsibility and authority. Beyond that, Lost Lake's isolated setting proved the perfect place for David to try out new ideas and refine old ones—if he could stay in camp long enough. One year, David's fellow campers blew a hole in the top of their communal tent when they accidentally ignited the stockpile of powdered magnesium he had brought to make fireworks. David was expelled. He was allowed back another year until, following in the late Harry Bennett's footsteps, he whipped up a batch of moonshine using a yeast-based recipe for ethanol, the basic component of distilled liquor, which he found in the *Golden Book*. David himself was a teetotaler. He made the moonshine at the request of a pair of older camp counselors, who had promised to move him off kitchen duty in exchange for the homemade booze.

Despite all the rewards of scouting, David from time to time gave serious thought to dropping out. It was partly due to the fact that troop meetings and activities cut into his time for scientific recreation. The other reason was that he was becoming embarrassed to belong to a group that many high school kids wrote off as a collection of dorks and losers. One day, he wore his uniform to school to give a presentation about scouting. He brought in awards he had won, talked about hiking and canoeing, and spoke earnestly about the importance of earning the rank of patrol leader. But at the end of the presentation, some of the kids teased him and asked questions designed to embarrass him (such as how best to go to the bathroom in the woods). "They made me feel like a jerk," David remembered. "I thought maybe belonging to the scouts just made me look stupid."

David talked to his girlfriend, Heather, about the scouts. She thought it was a perfectly honorable pastime and urged him to ignore the other kids. His father was even more adamant on the subject. When David raised the idea of dropping out of the scouts, Ken tried to talk him out of it. He saw scouting as an antidote to David's chemistry work, which to him represented a breakdown in discipline. Not recognizing his son's one great talent, he didn't encourage his scientific work, find him a mentor, or try to get him into an organized program for young scientists. Instead, Ken viewed scouting as a means of keeping David away from the one thing he was passionate about.

Furthermore, Ken was determined that David would accomplish a goal that he didn't himself achieve as a child: being an Eagle Scout. For Ken, who looked back upon his scouting days as a time of great hope, achievement, and belonging, attaining Eagle status was as worthy an honor as taking home the medal for the Olympic decathlon, and it always gnawed at him that he had fallen short. One night, when David was fourteen, Ken sat his son down at the kitchen table after the dinner dishes had been cleared away and awkwardly opened a rare heart-to-heart discussion.

"David," he began, pulling out a pouch filled with merit badges he'd earned in his failed quest for the Eagle, "this is the highest honor in scouting. I came up short, but you're my second chance." Partly out of his own desire and partly to get his dad off his back, David decided to stay the course.

Only about 2 percent of scouts attain Eagle status. Those who do are honored at an awards ceremony, where they are presented with letters from the president and their home-state members of Congress, as well as with a flag that has been flown over the U.S. Capitol. How does one join the eagle's nest? In addition to showing "scout spirit," Eagle Scouts must earn twenty-one merit badges, which they receive after completing tasks specific to more than one hundred possible topics.

Eleven of the twenty-one merit badges are mandatory, such as those for family living ("Inspect your home and grounds. List any dangers or lack of security seen") and citizenship ("Explain why you should respect your country's flag, repeat from memory the Pledge of Allegiance, and tell about two things you have done that will help law-enforcement agencies"). The scout is free to select the other ten merit-badge subjects from choices ranging from agribusiness to dog care to metalwork to stamp collecting. The work on each is reviewed by a counselor who tests the scout on his knowledge before awarding the badge.

David had completed most of the mandatory merit badges during his first few years of scouting, and by midway through his freshmen year at Chippewa Valley he was working on the optional components. One of his first choices was the merit badge in chemistry. The requirements included making a list of ten chemicals found in the home and describing their uses, showing how baking soda neutralizes an acid solution, and testing foods for starch and protein.

For a kid with David's scientific smarts, this was basic stuff. The real lessons he absorbed from these somewhat tedious tasks had nothing to do with chemistry or other subjects he later studied. Far more important to his future was the subtly delivered dose of political indoctrination he received from his merit-badge education.

The Boy Scouts have always claimed to be apolitical, but the group has had a decidedly right-wing character since Lieutenant General Robert Baden-Powell founded it, in England, in 1908. Robert MacDonald, author of a history of scouting called *Sons of the Empire*, says Baden-Powell was haunted by the collapse of the British empire as Japan, Germany, and the United States emerged on the world stage. A war hero whose military career included tours of duty in Africa, India, and Afghanistan, he believed Britain's fall was spurred by the decline of its citizenry, especially the working-class boys who arose after the industrial revolution and who smoked, loitered on street corners, and otherwise led undisciplined and aimless lives. To an old soldier like Baden-Powell, who liked to cite David Lloyd George's dictum "You cannot maintain an A-1 Empire on C-3 men," the solution was obvious: He would create an organization to stiffen the fiber of the nation's youngsters and turn them into a collective national ideological guard.

Scouting placed no restrictions on membership, and its political message was to be delivered with a great deal of discretion, so as not to frighten away any boys infected with Bolshevik or union beliefs. Hence, scouting would not overtly embrace political parties or issues (even if its biases were clear) but instead promote principles and behavior that were inherently conservative. In this regard, nothing was more important than discipline and unthinking loyalty to any and all authority—parents, teachers, employers, and political leaders.

Baden-Powell believed "a dull lad who can obey orders is better than a sharp one who cannot." (In other words, David was Baden-Powell's worst nightmare.)

Scouting for Boys, the hugely popular book Baden-Powell authored in 1908, laid down the Scout Law, which the new organization treated with the same solemnity that others hold for the Ten Commandments. In addition to calling for boys to be courteous ("especially to women and children and old people and invalids"), cheerful, and thrifty, it also demanded unswerving loyalty to "the King, and to his officers, and to country." The seventh of the Scout Law's nine rules states that this same degree of obedience is due to scouting's own leaders. "A Scout obeys orders of his patrol leader or Scoutmaster without question," it says. "Even if he gets an order he does not like he must do as soldiers and sailors do, he must carry it out all the same."

But discipline and patriotism alone would not be enough to save the tottering British empire. Baden-Powell was equally intent on wiping out a nefarious scourge that was eating away at the empire's vital parts: masturbation, or "self-abuse," which invariably produced "weakness of heart and head, and, if persisted in, idiocy and lunacy."

According to *Scouting for Boys*, "The training of the boy would be incomplete if it did not contain some clear and plain-spoken instructions on the subject of continence. The prudish mystery with which we have come to veil this important question is doing incalculable harm." As noted by Michael Rosenthal in his book *The Character Factory: Baden-Powell and the Origins of the Boy Scout*

Movement, the general never went so far as to describe exactly what he meant by the term *continence*, though he made it clear that the practice of masturbation was to be avoided at all costs.

With temptation always nearby, Baden-Powell's recipe for success was self-discipline. He urged scouts to avoid eating rich foods or sleeping on their backs in soft beds with too many blankets, both practices that could swiftly lead to ruin. For the compulsive "self-abuser," Baden-Powell recommended washing in cold water "to still the impulse" and, if all else failed, speaking to a scout officer "who can then advise him what to do."

England's social and political elite swiftly embraced scouting. H. G. Wells wrote of scouting, "There suddenly appeared in my world a new sort of little boy—a most agreeable development of the slouching, cunning, cigarette-smoking, town-bred youngster—a small boy in a khaki hat, with bare knees and athletic bearing, earnestly engaged in wholesome and invigorating games." The organization received a royal charter, and Winston Churchill later wrote in *Great Contemporaries* that Baden-Powell had bequeathed to England "an institution and an inspiration characteristic of the essence of British genius. It is difficult to exaggerate the moral and mental health which our nation has derived from this profound and simple conception." It was such a success that a Girl Scouts organization was launched in 1912, four years after the Boy Scouts debuted.

Not everyone was so enamored of the scouts. Critics derided the group as jingoistic and militaristic, as well as obsessed with rules and regulations. In 1909, P. G. Wodehouse lampooned the new organization in a short novel called *The Swoop!* It chronicles the dauntless deeds of Clarence MacAndrew Chugwater, the Boy

Scout who saved England from invasion by eight foreign armies headed by Germany and Russia, the latter under Grand Duke Vodkakoff, with additional contingents led by the Mad Mullah and "a boisterous band of Young Turks." The heroic Chugwater had perfectly mastered the art of scouting: "He could low like a bull. He could gurgle like a wood-pigeon. He could imitate the cry of the turnip in order to deceive rabbits. He could smile and whistle simultaneously in accordance with Rule 8. . . . He could spoor, fell trees, tell the character from the boot-sole, and fling the squaler." In the end, Chugwater and his scout troops—the Chinchilla Kittens, the Bongos, the Zebras, the Iguanodons, the Welsh Rabbits, and the Snapping Turtles—used catapults and hockey sticks to repel the foreign invaders.

Despite its occasional critics, scouting continued to thrive. During the 1920s and 1930s, vanquishing the menace of communism replaced stamping out masturbation as the focus of Baden-Powell's energies. He was far more indulgent of the fascist movements then sweeping across Europe, most notably Mussolini's Italy and Hitler's Germany, and found much about their organizations to admire. In articles for scouting publications, he stated that both dictators "realize that to be strong the nation must be united in patriotism," and that Hitler and Mussolini had "done wonders in resuscitating their people to stand as nations." (As late as 1937—two years before the outbreak of World War II—Baden-Powell sought to establish formal ties between the scouts and the Hitler Youth movement, which was modeled on scouting. Only a stiff rebuff from scouting's International Committee stopped him.)

Scouting had reached America's shores in 1910. According

to legend, William Boyce, an American businessman visiting London, was lost in a thick fog when a young man in uniform offered his assistance. The youngster not only led Boyce to his destination but refused a tip, explaining, "No, sir, I am a scout. Scouts do not accept tips for courtesies." Boyce wanted to know more about the organization, so the boy waited while he completed his business and then led him to the nearby office of Baden-Powell.

Upon returning home, Boyce rounded up a group of sponsors and soon incorporated the Boy Scouts of America. Scouting in the United States displayed roughly the same sensibility as it did in England (though Baden-Powell's fixation with stamping out self-abuse was thankfully absent). Its early champions were Progressives—political reformers who, like Baden-Powell, feared that industrialization and urbanization were eroding their nation's moral fiber. As one of scouting's early American champions put it, " 'City rot' has worked evil in the nation."

By 1914, scouting had a membership of one hundred thousand. That figure topped one million by the 1940s, and six million at the turn of the millennium. Scouting's most prominent graduates include President Gerald Ford, a former assistant scoutmaster who, while in office, appeared in a TV and print campaign for the scouts wearing the group's trademark red neckerchief; home-run champ Hank Aaron; actor Jimmy Stewart; astronaut Neil Armstrong; and film director Steven Spielberg, who subsequently wrote the requirements for the scout's merit badge in cinematography and said that a missed scout campout was the indirect inspiration for one of the most successful films of all time. Knowing that his buddies would be sitting around the campfire telling ghost stories, Spielberg, alone and bored at home, concocted one of his own. Many years later, he turned the basic outline of that story into the script for *E.T. the Extra-Terrestrial.*

As in England, American scouting has had its critics. Like Wodehouse, these detractors poked fun at the scouts' earnest, high-minded sense of purpose and conservative ethos. In 1953, musician Tom Lehrer poked fun at the group with his satirical song, "Be Prepared":

> *Be prepared*
> *That's the Boy Scouts' marching song.*
> *Be prepared*
> *As through life you march along.*
> *Be prepared to hold your liquor pretty well.*
> *Don't write naughty words on walls if you can't spell.*
> *Be prepared*
> *To hide that pack of cigarettes.*
> *Don't make book*
> *If you cannot cover bets.*
> *Keep those reefers hidden where you're sure*
> *That they will not be found,*
> *And be careful not to smoke them*
> *When the scoutmaster's around,*
> *For he only will insist that they be shared.*
> *Be prepared!*

Along with the other members of Troop 371, David imbibed the Boy Scouts' highly conservative orientation. Of course, he was too fixated with science to openly embrace politics and too much of a reckless loner to take seriously scouting's dogma of obedience and virtue. As he did with schoolwork, David heeded what he found interesting and useful and ignored everything else. One aspect of scouting that he unquestioningly absorbed was the group's long-standing endorsement of the nuclear-power industry.

The scouts worked closely with the Atomic Energy Commission (AEC), which had been created in 1946 and charged with promoting atomic power while protecting the public and the environment. In practice, the AEC focused almost exclusively on the promotional goal, producing brochures, TV shows, and traveling exhibits designed to sell the new technology to the public. The commission's film library included *The Atom and Eve*, which showed how nuclear energy would liberate women by powering toasters, hair dryers, and washing machines.

In a further effort to gain public acceptance for the industry, the AEC united itself with the irreproachable Boy Scouts organization and helped develop requirements for an atomic-energy merit badge. In November 1963, the first thirty-four scouts to qualify received their award from AEC chairman Glenn Seaborg at the annual meeting of the American Nuclear Society and the Atomic Industrial Forum in New York City. Seaborg, who was born and raised in the small Michigan town of Ishpeming, was himself a former Eagle Scout; he fulfilled his scout oath to do his duty for God and country as the leader of the Manhattan

Project team that in 1941 discovered plutonium, which was soon rushed into service in the bomb that obliterated Nagasaki. By 1967, some fifteen thousand scouts had qualified for the award, and news of their growing numbers had become a staple of the AEC's yearly list of achievements sent to Congress.

So shameless and enduring was its shilling for nuclear power that the scouts later helped the industry turn the partial-core meltdown and mass evacuation at Three Mile Island into a marketing opportunity. Beginning in 1984, just five years after the accident, the plant's then operator, General Public Utilities, allowed local Boy Scout units to visit Three Mile Island to meet some of the requirements for the atomic-energy merit badge. (TMI's Unit Two has been mothballed ever since the accident occurred. Unit One remains up and running.) A *New York Times* account of an early session reported that scouts had "donned hard hats, walked past pipes that had radiation tags and gathered in the control room where the nation's worst commercial nuclear accident [had] unfolded."

The kids performed practical exercises, such as using a Geiger counter to find a hidden radioactive "source"—for example, a coffee cup with a slightly radioactive glaze—and got plenty of handouts to take home. One, a booklet called *Nuclear Energy Facts: Questions and Answers* prepared by the American Nuclear Society, argued that the vibrancy of American democracy is best gauged in kilowatt-hours. "How important is it to have more energy—and more electricity?" read question number one of the booklet. "Very important," came the reply. "If we limit the amount of energy we have, we lose our freedom and our democratic society."

Before heading off of Three Mile Island, scouts watched videos and films like Walt Disney's *The Atom: A Closer Look*, an updated version of Disney's 1956 classic *Our Friend the Atom*. The overwhelmingly pro-atom movie, which the American Nuclear Society delivers free to "needy" schools, borrows heavily from the story of the fisherman and the genie in the *Arabian Nights*. In Disney's retelling, the fisherman opens a bottle containing a hideous genie, who is furious about having been trapped for so long inside the bottle. He is about to do away with the fisherman when he is tricked back into the bottle by him. After the genie promises to behave—and to grant him three wishes—the fisherman relents and lets it back out. Now the genie is the old salt's faithful servant.

A book that Disney produced to accompany the 1956 film opens with an illustration of a nuclear blast destroying a city and closes with one of a futuristic, sparkling city floating in the clouds. In between, author Heinz Haber spells out the moral of the story, in the unlikely event that anyone missed it. Like the fisherman, atomic scientists had released a terrible genie, which might destroy all humanity "with the most cruel forms of death: death from searing heat, from the forces of a fearful blast, or from subtly dangerous radiation." Thankfully, mankind possesses the scientific know-how to turn the genie's might to peaceful and useful channels. The solution to controlling the atomic genie was the nuclear reactor, which would magically transform "cruel forms of death" into power, food, health, and peace "for the good of man."

There's a curious and dark underside to this tale. During World War II, Haber was an astrophysicist who worked with

Wernher von Braun, the German missile designer whose V-2 rocket terrorized London and other British cities. An SS major who headed weapons research at the Peenemunde complex— where slave laborers were starved, beaten, and worked to death— von Braun could have ended up in the docket at Nuremberg like other high-ranking Nazis. But at war's end, the U.S. Army was anxious to plumb German scientific expertise in order to improve America's weaponry. Under the top-secret Operation Paperclip, the army smuggled von Braun, Haber, and more than one hundred other Peenemunde scientists to the United States. Von Braun went on to design the army's Jupiter rocket and in 1970 was appointed deputy assistant director of planning for NASA. (His metamorphosis from Nazi scientist to U.S. space hero inspired Tom Lehrer to sing of NASA's man, "Don't say that he's hypocritical,/Say rather that he's apolitical./'Once the rockets are up, who cares where they come down?/That's not my department,' says Wernher von Braun.") Haber, who died in 1976, went to work for the air force and founded the space-medicine department at Randolph Air Force Base in Texas.

In addition to working on *Our Friend the Atom,* Haber, joined by von Braun, collaborated on a number of other Disney projects. The former Nazis helped consult on Disneyland's Tomorrowland section and served as technical advisers on three space-related TV films in the 1950s that the company produced to promote the new theme park. The films were a huge success, with the first one watched by forty-two million people. Von Braun himself appeared in a segment that successfully foresaw the moon launch and the space shuttle, though it may have erred in prophesying an atomic-powered spaceship journey to Mars.

Soon after his dad convinced him to stick with scouting, David decided to pursue the merit badge in atomic energy. Like the choice of the chemistry badge, it was a natural for David, though for most kids it would have been an unusual one. It was 1991—David was fourteen—and popular support for nuclear power had been waning for at least a decade.

Too many events had polluted the public imagination. In 1979, the Three Mile Island plant near Harrisburg, Pennsylvania, came within hours of a full-scale meltdown. All that averted disaster was the arrival of a new shift supervisor, who came on duty as panic was spreading in the control room. He discovered that the level of cooling water was too low because thousands of gallons had poured through a valve that was stuck open. Meanwhile, the core was badly damaged, as were one third of the plant's fuel rods, and radiation levels above the facility reached three thousand millirems per hour, three hundred times higher than the release from an average X ray. Pennsylvania officials considered a mass evacuation of the region's six hundred thousand people, but ultimately settled for advising pregnant women and young children living downwind of the plant to leave the area.

No one died as a direct result of the accident at TMI, but the human cost is still today the subject of scientific debate, with some researchers suggesting that people who lived close to the plant suffered an increased incidence of lung cancer.

In 1986, seven years after the accident at TMI, there was a far worse catastrophe at the Chernobyl plant near Kiev in the Ukraine. There, a reactor blew up, and a cloud of radioactive strontium, cesium, iodine, and plutonium spread as far away as

the Mediterranean. Thirty people died in the immediate
aftermath of the accident, but the fallout is believed to have
ultimately killed thousands of people and poisoned hundreds of
thousands more. Many of the victims fell ill years after the
disaster from eating produce and drinking cows' milk tainted by
radiation. The financial consequences of Chernobyl were equally
devastating. According to *The Wall Street Journal*, the accident
cost the Soviet Union more than three times the total benefits
that accrued from the operation of every Russian nuclear-power
plant between 1954 and 1990.

At the time David began pursuing the atomic-energy merit
badge, *The Simpsons* had just become one of the most popular TV
shows, and its satirical but potent antinuclear satire garnered
huge laughs from an audience that had become used to being
skeptical about nuclear power. Homer, the family patriarch, was
a nitwitted nuclear-plant safety inspector who slept and ate
doughnuts on the job. The show's villain was Montgomery
Burns, the greedy plant owner who offered bribes to inspectors
who discovered nuclear waste sloshing about and scoffed at
concerns about the dangers of radiation, even when three-eyed
mutant fish began appearing in nearby ponds.

Curiously, David was a big fan of *The Simpsons.* He and his
mother would sit in front of the TV and laugh uproariously when
Homer accidentally spilled a bucket of glowing radioactive waste
or absentmindedly shook a nuclear fuel rod out of his shirt.
Nonetheless, David remained too much the eager beaver to take
to heart the cynical bite of *The Simpsons* or to dwell for too long
on the accidents at Three Mile Island and Chernobyl. He must
have been among the very few—innocent or uninformed—who

still held a 1950s-era love of nukes. Being intrigued by all things nuclear, he enthusiastically embarked on his atomic-energy merit-badge quest with the help of a volunteer scout counselor from the local community.

During meetings at his own home, the counselor showed David glow-in-the-dark radium clocks—which he told him could be found at junkyards—and other mildly radioactive materials, taught him to use a Geiger counter—something David would later turn to his own unique ends—and generally talked up nuclear energy. The counselor had a personal library that was well stocked with texts about chemistry and nuclear power, and David borrowed books that further fueled his imagination. "He was very gung ho—almost as much as me," David said of his counselor. "Every time I saw him I came away wanting to learn more about nuclear power. I thought he was an adult who had a great life. He knew stuff and did stuff that was exactly what I wanted to be doing."

In many ways, the education in radioactivity that David received in Clinton Township Troop 371 was virtually identical to the one dished up to scouts at Three Mile Island. Like the *Golden Book*, David's atomic-energy merit-badge pamphlet resonated with the ethos of the 1950s and 1960s, from its technological triumphalism to the picture of a nuclear-plant executive whose thick, striped tie and suit made him look like a character out of *Father Knows Best*. Even the very name of the merit badge was hopelessly out-of-date. In popular lingo, the term *atomic energy* had long been supplanted by *nuclear power*. The former quaintly harked back to a bygone time when the atom represented one of science's greatest hopes, not a force that engendered one of the most sweeping protest movements of modern times.

Such was the pronuke slant of the pamphlet that it comes as no surprise that it was authored by a group of nuclear-power advocates that included Westinghouse Electric, the American Nuclear Society, and the Edison Electric Institute, a trade group of private utility operators. A number of individuals also helped in the preparation of the pamphlet, including Dr. Warren Witzig, who is identified in the acknowledgments as the former head of the Nuclear Engineering Department at Penn State University. Omitted is the fact that Witzig was also a longtime board member of General Public Utilities, the firm that operated the Three Mile Island plant at the time of the accident.

The pamphlet instructed scouts that understanding atomic energy starts with the recognition that its critics descend from a long line of history's naysayers and malcontents: "In the late 1800s, people feared electricity. It was just starting and people did not understand what it could do. . . . In 1890 at Auburn prison in New York, Willie Kemmler died in an electric chair. This was supposed to prove how deadly evil alternating current electricity was."

The Boy Scouts systematically whitewashed the many problems encountered by nuclear power. For example, the pamphlet conceded that "a small amount of radioisotopes were released" during the accident at Three Mile Island but suggested that the whole episode was little more than a case of failed public relations. "Because of poor communications many people did not understand what was happening," the pamphlet stated, though it did not divulge that people didn't know what was happening because the plant's operators systematically lied to the press and public and tried to cover up the radiation releases altogether. David's pamphlet (which was published in 1983) didn't mention at all the far more serious disaster at the Chernobyl nuclear-power plant.

For someone already performing fairly sophisticated if reckless experiments, the requirements for the merit badge were pathetically easy. David drew and colored the radiation hazard symbol, used Ping-Pong balls, clay, and marshmallows (representing neutrons, protons, and electrons) to make models of various atomic elements, and, with the help of a chart found in the pamphlet, calculated his annual dose of radiation from natural and man-made sources. For the latest year, it had come to 308 millirems, about 50 percent higher than the national average. The spike was largely due to a dose of 98 millirems David received when getting a skull X ray after his misadventure with red phosphorus.

As a final set of tasks, David built a Geiger counter from a kit, visited a hospital radiology unit to learn about the medical uses of radioisotopes, and constructed a model nuclear reactor originally designed by Bob LeCompte, a former Boy Scout who had subsequently moved on to the AEC. The model was built of a juice can (for the core), plastic pill bottles, coat hangers, soda straws, kitchen matches, and rubber bands. The latter were used to make a scram spring, simulating the emergency shutoff system of a real reactor. David was so rigorous in fulfilling the requirements that his counselor joked that he deserved *three* merit badges in atomic energy.

David's Boy Scout education profoundly influenced his
nuclear dreams. The projects he was required to complete may
have been goofy, yet their very simplicity allowed him to
envision future projects on a far bigger and more sophisticated
scale. David discovered with joy and amazement that getting his
hands on radioactive materials—even ones used by the Curies!—
might not be as difficult as it seemed. The pamphlet informed
him that an isotope of polonium—an element that the Curies had
discovered and named—could be found in electrostatic brushes
used to clean film negatives. (Polonium produces a charge that
attracts dust.) Americium-241, he discovered, is used in smoke
detectors because the alpha radiation it emits is easily stopped,
even by a wisp of smoke. (When the alpha particles are blocked
in the air, a sensor triggers the detector's alarm.) And his
counselor had pointed to junkyards as a source of radium.
All this and more he stored and catalogued in his head for
future use.

Perhaps because both groups are polemical and have
attracted strong criticism, the Boy Scouts and the nuclear
fraternity almost demand blind allegiance from their members.
You're in or you're out; there's no room for criticism or doubt.
That was certainly the case for David; years later, when
answering questions on the subject, he sounded as if he were
regurgitating lessons learned from the merit-badge pamphlet—
no surprise, since by his own estimate he had read it hundreds of
times. "I knew about the accident at Three Mile Island, but it
really didn't sound like the problems were all that serious," he
said. "Anyway, I thought there would always be new
developments with nuclear energy and that the plants would

keep getting safer and more efficient." David was scornful of the powerful antinuclear movement that burgeoned during his youth. It was filled with people who had watched too many bad science-fiction movies and were petrified of their own shadows—just like those people from an earlier era who had feared electricity. "I thought those groups were made up of people who were ignorant, who were just afraid of the unknown," he recalled.

David was awarded his atomic-energy merit badge on May 9, 1991, five months shy of his fifteenth birthday. He still needed to earn about another ten merit badges to become an Eagle Scout, but the Boy Scouts had already helped stir bigger hopes.

CHAPTER 5

Stalking the Periodic Table:
Elements, My Dear Watson

Invention comes to those . . . who can rebound from the continual snickering behind their backs when things go wrong. To be a successful inventor, one needs a very thick skin.

—IRA FLATOW, THEY ALL LAUGHED, 1992

For a long while, David didn't have a master plan or even a definite goal for his scientific research. His interests had been all over the map, including old standbys such as tanning and explosives but also more exotic pursuits such as DNA and the healing power of antioxidants. Earning the merit badge in atomic energy had served to focus his thoughts. Late in his sophomore year at Chippewa Valley High School, halfway between his fifteenth and sixteenth birthdays, David's inventive energy began to take an increasingly distinct if still not wholly defined direction.

One of the first people to get a whiff of what David was up to was his physics teacher, Ken Gherardini. "His dream in life was to collect a sample of every element on the periodic table of elements," he later recalled, in a voice mixed with amusement and quiet mystification. "I don't know about you, but my dream at that age was to buy a car."

David's was indeed a remarkable and imaginative pursuit. The periodic table lists the elements in order of increasing atomic number (the number of protons in the nucleus of an atom). Most of the elements are common and benign, such as calcium, nickel, copper, gold, and iron. The final thirty-four on the periodic table, from number 84 (polonium) to number 114 (identified in 1999 and as yet unnamed), are more exotic—many of them man-made elements not found outside the laboratory. All thirty-four, though, share a common trait: They are radioactive. Here we find uranium, radium, plutonium, and thorium, as well as a number of elements named after the nuclear pioneers: curium, einsteinium, fermium, seaborgium, and hahnium.

David didn't want to construct a collection to keep locked in a glass display case; he wanted to conduct experiments. "I was becoming more and more interested in radioactivity and atomic power," he later said. "I didn't know exactly what I would need or even what I would do, but I wanted stockpiles of as many elements as possible, just in case."

David would sometimes come to his physics class with talk about the wild experiments he'd conducted during the weekend, even hinting that he was engaged in radioactive research. Gherardini, understandably, never took his claims seriously. He figured David was simply trying to conceal the facts that no one wanted to spend time with him and he didn't have anything to do on weekends. As far as gathering up the elements, Gherardini assumed David wouldn't actually attempt to obtain *everything* on the periodic table; surely he intended to concentrate on simple items—nothing more obscure than helium or neon—and draw the line before reaching the upper end of the table. As was always the case, David's aspirations were far more ambitious than the adults in his life could imagine.

Unlike many of his idols, David had no financial support from the state, no laboratory save for a musty shed, and no proper instruments or safety devices. This last issue rarely troubled David, and not merely because he, like the typical teenager, gave little thought to his own health or mortality. As time went on, David's experiments became more and more of a refuge. Chemistry was utterly predictable; if you conducted a test properly, you could always count on the right result. With people, David found, that frequently wasn't the case.

Of course, he did have Heather, with whom he maintained a hot and heavy romance. Completely smitten and perhaps amazed at his luck in getting such a desirable girl to like him back, David tried his best to be a normal boyfriend, despite his preferences for isolation and experiment. Every Friday, he picked up Heather at her house at 7:00 P.M. punctually. They'd head for a party, a movie, the Putt-Putt course, or, more often than not, CJ Barrymore's, an arcade that featured video games, baseball batting cages, go-carts, and pinball machines. Though distracted by science, David was devoted to Heather. One day, he was sure, they would marry.

Still, he didn't have the emotional sophistication to explain to her—or even to fully understand himself—just what about his home life distressed him or why he sought comfort in quantifiable reactions. Indeed, with the exception of his relationship with Heather, David's emotional life remained a mess. His mother often seemed on the verge of unraveling and, though she still offered her son love, didn't provide much in the way of parenting or stability. She'd fight with Michael and go to her mother's house, then fight with her mother and spend nights sleeping in her Mercury Cougar, parked in the driveway. "I kept hoping things would change, that she'd stop drinking and get better, but it never happened," David said. Meanwhile, his absent and detached father was incapable of developing an intimate relationship with David, nor did he have the patience or temperament to draw him out. Ken would sometimes find time for a scouting campout with his son—that offered safety in numbers—but not for a weekend alone. Just like David, Ken was engrossed in his work, which helps explain why his son was able to pursue sophisticated and dangerous scientific research in relative peace.

Whether he was mixing chemicals in his basement lab or isolating enzymes at the potting shed, science was David's primary means of coping with the world, building his confidence, and showing off. Sure, he understood something about the dangers of radioactivity, but David wouldn't let that stop his nuclear research. After all, where would that have left him? He'd no longer be an intrepid young scientist working on cutting-edge research but, once again, an insecure, unhappy teenager with few friends, who had a hard time fitting in at home or at school.

David's hunt for the elements began where Gherardini thought it would end, namely with easy-to-find items like copper and gold. He got sulfur from a rock collection, carbon from burnt coal, and tungsten plucked from lightbulb filaments. David mounted and carefully labeled his finds on a sheet of poster board that he hung on the wall of the basement.

Before long, though, David was seeking more dangerous elements. He received some help from a high school friend who had a job at the Macomb Community College laboratory and would occasionally filch a sample of an item David needed. (Curiously, David never considered seeking employment at the lab. "I feared legal repercussions through the misuse of authority," he explained—this being a stilted way of saying that he felt that the urge to steal the lab blind would be irresistible if he worked there.)

This friend got David one of his early acquisitions, phosphorus, number 15 on the periodic table. The element was contained in six light yellow candle-shaped sticks, which were packed in a can of oil because phosphorus ignites upon contact with air. David and his friend waited for the sky to darken one evening and drove to a thinly trafficked stretch of Canal Road in Clinton Township. Using tongs to handle the phosphorus, David flung one of the sticks on the pavement and watched it glow a vibrant green, then spontaneously ignite. David threw water on the phosphorus, which caused the flames to erupt into a bonfire-size blaze. At the sound of approaching fire engines, he and his buddy hightailed it home.

Sue Young, David's chemistry teacher, was vaguely aware of David's pursuit of the elements on the periodic table. As an adult, David would credit Young with being a mentor, saying her class was a great inspiration to him. The admiration, though, wasn't entirely mutual. Young thought David had a great aptitude for chemistry but believed he was immature, had poor study habits, and lacked the discipline needed to be a scientist. With his talents, David should have been a shining star; instead, he daydreamed about space travel and nuclear energy and wallowed in academic mediocrity.

David once discussed his periodic-table project with Young, but she, like Gherardini, assumed he was after only the basic elements. Young soon forgot about the entire conversation. Some time later, Chippewa Valley hosted an evening of parent-teacher conferences. As Young sat at her desk talking with one parent, she noticed a man waiting his turn who was anxiously peering into a brown paper bag that he held in his lap. After concluding with the first parent, she signaled the man—Ken Hahn—to join her. Almost before he hit the chair, Ken pulled out of the bag a small can that his wife had found recently on a shelf in David's basement laboratory. David hadn't confessed to what was inside but insisted that the substance was absolutely harmless and that there was no cause for alarm. Now Ken wanted to know if Young could identify the mysterious matter and tell him if it was dangerous to keep around the house.

Young opened the container and found what looked like a

stick of white, waxy butter floating in oil, which she recognized as a large chunk of sodium, element 11 on the periodic table. Young was astounded. Sodium is highly explosive; if exposed to air or water, it emits a blast of white-hot heat and explodes. Young was especially perplexed as to how a fifteen-year-old student had come to possess sodium; even she couldn't purchase it because it was far too dangerous to handle in a school setting.

The consequences could have been deadly if Ken or David—who had gotten the sodium from the friend who had supplied the phosphorus—had taken the sample out of the oil. A few years after Ken's meeting with Young, a University of Massachusetts student in Boston dropped a small container of sodium while cleaning out a laboratory refrigerator. The container popped open, splashing into a pool of water, and the sodium ignited. The student suffered severe burns and was rushed to the hospital along with eighteen others, some of whom had inhaled poison gas. The campus had to be evacuated, and the laboratory suffered two hundred thousand dollars' worth of damages.

But life with the Hahns was routinely bizarre. In most households, an adolescent caught handling something as explosive as sodium would have been in serious trouble; the affair itself would become one of those stories that are told and retold so often that they become a part of family mythology. For the Hahns, though, it was run-of-the-mill and soon passed entirely from memory. David and Ken remembered the story only after Young mentioned it to me during a phone interview. Why, yes, we did have a little problem with sodium, Ken said the next time I spoke with him, his memory now refreshed. He remembered Young's eyes lighting up—probably more from fear than from excitement—when she realized she was holding a can of sodium. At her suggestion, Ken turned the container over to the fire department for disposal.

When Ken confronted David, his son hit the roof. "Dad, that's my area. I respect your stuff, and I want you to respect my stuff," he said.

"Look, Dave, I'll try to respect your privacy, but you can't bring stuff into the house that can blow us up," Ken replied, drawing a typically generous line in the sand.

By now, David's oddball reputation was spreading widely at Chippewa Valley, where he was now known among his few friends as Glow Boy. David took the teasing in stride. Whatever other kids might say, he knew that his research, though apparently aimless and haphazard, would one day lead to significant findings.

At lunchtime, David would burst into the cafeteria just itching to talk to someone, anyone, about a scientific dilemma or research obstacle. He frequently sought out Jeffrey Hungerford, who was a year ahead of him in school and also precocious when it came to the sciences, especially chemistry. What sort of reaction would you get if you mixed two substances together? he wanted to know. How would neutrons affect certain isotopes? How many elements were fissionable?

David began bringing to the lunchroom the Geiger counter he had assembled for his merit badge. Joking that there had to be some explanation for the dismal quality of Chippewa Valley cafeteria fare, the kids would test lunch trays for radioactive emissions. "The Geiger counter used to always click faster when David held it," Jeffrey recalled. "I used to think that he had something radioactive in his pockets, but, in retrospect, maybe he was radioactive himself."

Like his teachers, his schoolmates didn't pay much heed to David's wild talk. After all, plenty of adolescents have strange habits, and David's weren't ostensibly weirder or more destructive than those of other high school kids. He talked cryptically about pursuing a form of nuclear energy and frequently sported a burn, scar, or other injury, but who could possibly believe his claims? His enthusiasm was interpreted as bluster, and even those who knew the most about David's research figured he'd never obtain highly controlled radioactive materials let alone use them toward generating nuclear energy—it seemed beyond the capacity of a teenager in his backyard.

The following is a verbatim excerpt from a research paper David wrote during his junior year at Chippewa Valley. It discussed the vital importance of protecting the nation's environment, an inadvertently ironic topic given David's covert activities. The paper, clearly written without the aid of computerized spelling and grammar checks, opens:

There are more chemical and dump sites than anyone has ever expected. Sites are becoming overfiller to the brim. . . . Companies are dumping (swiftly) these chemicals directly onto the land or indirectly by streames. Other large companies are choking theirselves. The southern California the main "sulfur" mines are constantly being taken. The innate element is then converted into steaming hot sulfuric acid. The acid is eventually exported but millins of tons of acid are boiled into the atmosphere. This pouring of acids into the atmosphere is believed to cause acid rain. In Detroit, Michigan chemical waste are now being pushed into the old salt (NaCl) mines. If some kind of change does not occur America will definatly be in trouble.

No wonder no one believed David's wild tales. He was a student who could not spell *millions* but claimed nonetheless to be conducting advanced research in his backyard. David's academic mediocrity obscured the extraordinary talent he had in one area, and his seeming ordinariness proved to be an accidental yet effective cover for his research efforts. Doubtless, he preferred it this way. For beneath the blank exterior, his thoughts were bubbling away in ways that would later astonish those who knew him.

For his sixteenth birthday, Ken bought David a used brown Pontiac 6000. The car wasn't much to look at, but David loved it. He took some of his closely guarded savings and bought a car stereo system. He installed it himself and souped it up with an equalizer and reverberator that produced a bass effect that made the car shake. David had picked up a taste for pop music from his mother, and he cruised through the suburbs with Heather, listening to Laura Branigan, Stevie Nicks, and local Detroit heroes Diana Ross and Madonna.

When he wasn't with Heather, David was stepping up his search for as-yet-unattained elements. His success in obtaining sodium and phosphorus led him to grow more ambitious—and reckless. He was tired of fooling around with the elements at the lower end of the periodic table; he was ready to move on to some of the more exotic substances, especially numbers 84 and up.

Here David faced what at first glance appeared to be an insurmountable obstacle—namely, that the radioactive elements that so intrigued him were all tightly regulated by the federal government. But David had discovered a secret, which had been first revealed to him when he read in his Boy Scout materials about polonium and americium: Many household and consumer items contain radioactive elements. Perhaps they contained only small quantities and certainly not in a pure form, but David figured he could devise means of isolating and gathering radioactive elements from store-bought goods.

David needed expert advice to discover additional natural and commercial sources of radioactive materials. Gherardini and Young were quite knowledgeable about radioactivity, but he feared that his teachers would get suspicious if he asked too many pointed questions. It would be better, he decided, to consult out-of-town experts who didn't know him and to pretend that all of his questions were purely hypothetical.

And he knew just where to turn for help. The final page of his scout pamphlet contained a list of government agencies and industry groups that scouts could go to for additional information: the Department of Energy, the Nuclear Regulatory Commission, the American Nuclear Society, and the Edison Electric Institute.

David found a few other nuclear-related organizations on his own and began writing dozens of letters of inquiry, sometimes to multiple sources at the same organization.

Initially, he identified himself as a student seeking information for a school project, but it occurred to him that requests from a teacher might be treated with more respect. He came up with the idea of passing himself off as "Professor" David Hahn, an earnest, dedicated physics instructor at Chippewa Valley High School forever seeking ways to enrich his students' academic lives. Given that Professor Hahn would sometimes send daily requests for information, his contacts must have been greatly impressed, if not fairly irritated, by the enthusiasm he brought to the classroom.

David's letters didn't fool everyone, probably because Professor Hahn, even when consulting a dictionary, misspelled so many words and made so many elementary grammatical errors. Some letter recipients might also have found it odd that Chippewa Valley's star teacher never wrote on letterhead and sometimes requested information in sloppy, handwritten notes.

Still, David was sufficiently steeped in the discourse of nuclear engineering to con some government and industry experts into believing that he was a teacher and professional colleague. He got a reply to about one of every five letters he sent out.

Even then, he could never state his true intentions, so much of the information he received in response was of only marginal value. The American Nuclear Society sent Professor Hahn a teacher's guide called *Goin' Fission*, which included games such as a word search where students circle hidden terms like *fuel rod*, *breeder*, and *control rods*.

Yet much of what he received provided useful tips. *Dreams and Dragons*, another brochure sent by the American Nuclear

Society, was no more sophisticated than *Goin' Fission*, but it proffered one amazing piece of information. The mantle used in commercial gas lanterns—the silky bag that looks like a doll's stocking and conducts the flame—is coated with a compound containing thorium-232, which makes it glow especially brightly. A silver-white metal discovered in 1828 by Swedish chemist Jöns Jakob Berzelius and named after Thor, the Norse god of thunder, thorium is number 90 on the periodic table, two spots below uranium. It is intensely radioactive and has a half-life of fourteen billion years.

The Nuclear Regulatory Commission, too, proved to be a source of abundant information. The NRC was created in 1975 to succeed the Atomic Energy Commission and was every bit the industry lapdog that its predecessor agency had been. The NRC is a fee-based agency that gets its budget not from taxpayers but from the corporate plant owners, who are required by law to support it. Since the plant operators—among the biggest are Westinghouse Electric, General Electric, and the Southern Company—despise regulation and fees, they are forever lobbying to slash the NRC's budget. During the 1990s, the NRC's number of safety inspectors was slashed by 20 percent.

The NRC's most important contribution to David's nuclear quest was a list of commercial sources for many radioactive materials. This list was part of a large packet of background reading and was meant to show, reassuringly, that many industrial and household products contain small amounts of radioactive material. David, though, viewed it as a shopping list and guide. It wasn't possible for him to purchase all the items on the list—for example, industrial shipping containers made with trace amounts of uranium—but the list did supply several options that were more pragmatic, at least for someone with David's talents and perseverance. For example, tritium, a radioactive gas used to boost the power of nuclear weapons, is utilized in the manufacture of glow-in-the-dark gun and bow sights and to light exit signs on highways and in theaters.

He learned that hospitals carry cobalt-60 to treat cancer and that thorium is also found in certain ores. He had already known that uranium was contained in pitchblende and now read that it was once used in a glaze applied to orange-colored Fiesta dishes made in the 1930s. David also discovered that in the United States alone there are several million radioactive machines and tools in use. These tools contain isotopes, doubly encapsulated in stainless steel, that emit gamma rays or neutrons (cesium-137, among others, generates the former; a mixture of americium-241 and beryllium produces the latter). Both gamma rays and neutrons have penetrating properties and, acting like X rays, enable technicians to see through asphalt, concrete, steel, and other hard surfaces. Hence, radioactive tools allow engineers to check for cracks in bridges, airlines to inspect baggage, and bottlers to check fill levels.

The Record, a New Jersey newspaper, reports that exposure levels from commercial sources of radiation are generally minimal, but the NRC has recorded hundreds of cases in which Americans received doses higher than deemed safe by the federal government. A New Jersey teenager once dragged an exit sign out of a demolished building, broke open several tubes contained in it, and got hit with a blast of tritium gas. For two weeks, government workers scrubbed down his entire house and destroyed every single item from his room, from clothing to wall posters. In Texas, a cop picked up an odd purple object lying in the grass off the highway. It turned out to be a discarded atomic battery from a well-exploration tool, which gave him a dose fifty times higher than deemed safe by authorities. An NRC employee was strolling down the street in King of Prussia, Pennsylvania, when he found a radioactive gauge on the sidewalk.

By now, David had more than enough information to jump-start his research. "I kept getting more and more pumped up," he later said of these heady days of exploration and discovery. David might have recalled the Curies smashing their tons of Bohemian ore or Fermi with his atomic pile beneath the football stadium at the University of Chicago. He was not a vulnerable, isolated kid but another pioneer of science, searching for the key to new and exciting radioactive discoveries.

David now replaced his first Geiger counter, the one he'd made from a kit for his merit badge, with a more sophisticated model that he purchased from a mail-order house in Scottsdale, Arizona, and mounted it on the dashboard of his Pontiac. He assured his dad that he used the Geiger counter only to test natural sources of radiation, such as ores and rocks he found in the woods. So as not to provoke additional domestic doubts from Ken and Kathy, he began stashing chemicals and equipment for his experiments in the trunk of his car.

To further aid his radioactive scavenger hunt, David distributed a list of desired items to a few friends. Several agreed to help him, though they still didn't take his activities too seriously. "I thought the most he'd do was ruin any chance he had of having children," Andy Hungerford said glibly.

Despite the lack of faith displayed by his peers, David slowly yet methodically began to collect the materials on his shopping list. He did take some moderate new precautions before embarking on this phase of his hunt. He bought a charcoal-filter gas mask and "borrowed" an old lead-lined protective suit from a government civil-defense agency in Detroit, professedly for a demonstration he was preparing for his Boy Scout troop. (The agency apparently lost track of the suit and never asked for it back.) He wasn't always religious about using his protective gear, though, especially on hot days when the suit and gas mask made the stifling potting shed even more unbearable.

David's first triumph, a modest one, was isolating a sample of polonium, which he got by buying a few electrostatic brushes through the mail for about twenty dollars apiece. The dark brown camel-hair brushes were about three inches long. A stick-on label next to a thin aluminum bar on the plastic handle warned: "Radioactive: Polonium-210 inside." David donned a pair of dish-washing gloves and used a wire cutter to bend back the corner of the aluminum strip. With tweezers, he pulled out the tiny silver strip of polonium and dropped it into a vial.

Americium, which was first identified by Seaborg and three other scientists during the Manhattan Project, proved to be just as simple an acquisition. David got his first batch (but by no means his last) during a scouting trip to Lost Lake Summer Camp. While most of the boys were sneaking into the nearby Girl Scouts camp, David executed a blitzkrieg raid on several unoccupied cabins and liberated smoke detectors from the ceilings.

David wasn't sure where the americium was located, so he

wrote to a smoke-detector manufacturer, BRK Brands in Aurora, Illinois, in the pose of a student preparing a research paper. A customer-service representative named Beth Weber wrote back to explain that each smoke detector contains only a tiny amount of americium-241, which is sealed in a gold matrix to make sure that corrosion does not break it down and release it. Thanks to Beth's tip, David was able to extract the americium components by bending the outer casing of the matrix with a pair of pliers. Out popped a tiny silver disk, about half the size of the end of a thumbtack.

Buoyed by these early and easy coups, David's ambition soared: He now decided to go after thorium. Thorium was originally used to put the fluorescence in gas street lamps. Because it has a melting point of about 3,300 degrees centigrade, it is nowadays employed in the manufacture of airplane-engine parts that reach extremely high temperatures. Any individual or company possessing thorium must have a license from the NRC, and the NRC is stingy in doling them out. Beyond a few aerospace manufacturers and university labs, thorium is not generally found in commercial or academic settings.

David knew from *Dreams and Dragons* that thorium dioxide is found in gas-lantern mantles. Manufacturers say the lanterns emit only a low level of radiation, though they recommend that campers wash their hands after changing the mantle. On the other hand, researchers in Saudi Arabia, where gas lanterns are commonly used to light villages, have found that the mantles emit enough radioactivity to cause long-term biological damage.

David began contacting surplus stores that sold hunting and camping equipment. After a few dead ends, he found and bought a few dozen old lantern mantles from a shop called Ark Surplus, then reduced them to ash with a blowtorch in the potting shed.

For David, handling dangerous items was fascinating and gave him a real sense of power. Soon, he began carrying his radioactive finds to school, to show off. First, he brought his strips of polonium—wrapped in a packet of aluminum foil—and the plastic handle with the stick-on radioactive-warning label. The polonium wasn't much to look at, though, nor did it impress anyone. One kid he showed the strips to said they were probably all that David had and challenged him to bring in something else—if he really had it. The next day, David came prepared, carrying in his backpack a Geiger counter and a Ziploc bag that was one quarter full of thorium ash. He invited a group of five kids to come with him, and they slipped into an empty chemistry classroom. David placed the Baggie on a lab table and told the kids it was thorium.

"Oh, yeah," said one of the kids. "That's nothing but dirt."

It was exactly what David had expected. With a flourish, he pulled the Geiger counter from his backpack and urged the skeptic to test the Baggie. When the kids in the room heard the Geiger counter begin to click loudly, they no longer doubted David's claims. In fact, they were so worried about being irradiated that David had to calm them by explaining that thorium emits alpha particles, which don't pass through plastic. "A lot of the kids had always said I was full of bull, that I couldn't get stuff like thorium," he recalled with a sly grin. "You should have seen their faces when they heard the Geiger counter."

A few Chippewa Valley students became nervous about David's activities, especially after he displayed a burn on one arm that he said was caused by radiation. Some friends switched desks in classes so as not to sit next to him; others stopped hanging out with him at all. But given the informal yet rigid teen code of silence, which bars collaboration with adults, no one reported concerns about David's activities to parents, teachers, or any authority figure.

All the while, David was becoming more and more versed in the esoterica of nuclear physics. Based on what he could understand when David began riffing on his acquisitions and discoveries, Ken concluded that his son was exaggerating the scope of his research in order to attract attention. Still, he decided he should look more deeply into his son's activities. He took David to meet with a chemistry professor he knew at nearby Oakland University. The professor spent one hour with David and afterward told Ken that his son's theories about nuclear science were founded on a wealth of sophisticated knowledge. Ken was impressed—and also distressed. Up until then, he'd been under the impression that David was just a kid screwing around with things he couldn't understand. "We may have a problem here," he told Kathy that night. "David knows enough to be dangerous."

Other odd occurrences soon heightened his fears. First came the pill vials he found hidden in David's room, filled with something that looked like paint flakes. Then there were the letters and boxes that came to the house from government agencies and companies scattered across the country. But David convinced his father that all this was part of research he was doing for scouting projects or for school. Ken chose to take him at his word. "He's a clever kid, and he was always careful to make sure that I never found anything too incriminating," Ken later said by way of explaining his laissez-faire approach to parenting. "I never saw him turn green or glow in the dark. I was probably too easy on him."

Michael and Patty were equally indulgent of David's experimenting. Naturally, they thought it odd that he had taken to wearing a gas mask in the shed and would sometimes discard his clothing after working there until two in the morning, often by flashlight, but they chalked it up to their own limited educations. "I was suspicious for a while there," Michael said, "but Patty thought he looked cute."

Cute, perhaps, but more dangerous than ever, for David's ambitions were becoming increasingly grand—and reckless. No longer would nuclear models built from marshmallows, pill bottles, and Ping-Pong balls satisfy. As Barbara Auito, the scoutmaster's wife and Troop 371's treasurer, later put it: "The typical kid [interested in nuclear energy] would have gone to a doctor's office and asked about the X-ray machine. Dave had to go out and try to build a reactor."

CHAPTER 6

*How Nuclear Enthusiasts Planned to
Solve the Energy Crisis: A Brief History
of the Breeder Reactor*

Plutonium will be the fuel of the future [and its value] may
someday make it a logical contender to replace gold as the
standard of our monetary system.

—GLENN SEABORG, CHAIRMAN OF THE
ATOMIC ENERGY COMMISSION, 1968

A conventional nuclear-power plant operates in
essentially the same manner as a power plant that runs
on fossil fuels like coal or oil. Heat boils water to make steam,
which turns a turbine generator. "The only difference is the
source of the heat," explained *Atoms to Electricity*, a booklet put

out by the Department of Energy. "Whereas coal and oil are burned directly, a sustained atomic chain reaction called fission generates heat in a nuclear plant."

There are several dozen radioactive elements, but only a few isotopes with particularly unstable nuclei can undergo fission: uranium-235, plutonium-239, and uranium-233, the last two of which are man-made. Fission occurs when a neutron (or other high-speed particle) is fired at and penetrates the nucleus of an isotope.

For example, the nucleus of an atom of U-235, the most easily fissionable isotope because it has unusually large atoms that are highly volatile, contains 92 protons and 143 neutrons. When a U-235 atom is hit by and absorbs a neutron, it splits in two—fissions—and releases energy in the form of heat, as well as a few additional neutrons. These neutrons are in turn absorbed by other U-235 atoms, beginning the process again. In theory, you need to split only a single U-235 atom to create a chain reaction. The neutrons from the first split will continue to split other atoms, which will split more and more in geometric progression. If left unchecked, a chain reaction will lead to an uncontrolled meltdown.

Unlike its sister isotope, U-238 (92 protons and 146 neutrons)—which comprises over 99 percent of the content of pure uranium ore—is not itself fissionable. That's because it has three more neutrons in its nucleus than U-235 and tends to reflect additional neutrons rather than absorb them. U-238, though, is considered "fertile." As discovered by Glenn Seaborg, when bombarded with neutrons U-238 is transformed into U-239. The latter decays by emitting beta rays

and is transformed into a man-made radioactive element that
Seaborg and his team called neptunium—after the planet
Neptune, a name picked because the element contains in its
nucleus one more proton than uranium, which of course had
been named for Uranus, the planet next closest the sun.
Neptunium is unstable and spontaneously ejects a beta particle,
thereby forming plutonium (named for the planet Pluto), with
ninety-four protons. Plutonium, which is fissionable, can be
used as fuel for a nuclear reactor.

Nuclear-power plants are built to various designs, but the
core, where heat is produced, generally comprises four basic
components. The first is the fuel, usually pellets of uranium
encased in tubes. Firing neutrons at these fuel rods produces
fission. The second is the moderator, which is used to slow down
the neutrons. Manhattan Project scientists discovered that some
neutrons move at about seventeen million miles per hour, one
fortieth the speed of light. If they are "moderated" to about five
thousand miles per hour, they have a better chance of being
absorbed by another atom and provoking fission. Ordinary water
serves as an excellent moderator and does double duty in the
core by acting as the third component, the coolant, which
conveys the heat produced by fission and powers the turbines.
The fourth and final component of the core is the control rods,
which today are generally made of boron (number 5 on the
periodic table). Boron is capable of absorbing neutrons but is
not itself fissionable. When the control rods are pulled out, the
fission reaction increases. If they are inserted, they block the
neutrons, and the chain reaction slows to a halt.

Uranium is a hugely efficient fuel source. Thirty-three tons can produce as much energy as 2.3 million tons of coal, 10 million barrels of oil, or 64 billion cubic feet of natural gas. But just as a coal-fired plant requires regular replenishment of fuel, so does a conventional nuclear-power plant require fresh uranium to feed the reactor core. However, as explained in *Atoms to Electricity*—which David wrote away for and used as a primer—a breeder reactor in theory never needs new fuel once successfully up and running. As the pamphlet rhapsodized, "Imagine you have a car and begin a long drive. When you start, you have half a tank of gas. When you return home, instead of being nearly empty, your gas tank is full. A breeder reactor is like this magic car. A breeder reactor not only generates electricity, but it also produces new fuel. In fact, a breeder reactor can produce more fuel than it uses." In other words, just as medieval alchemists sought a means to transmute base metals into gold, nuclear engineers would transform plutonium and uranium into eternal light and power.

And just how does a breeder reactor achieve this miraculous feat? A breeder reactor is configured so that a "blanket" of U-238 surrounds a core of plutonium-239 fuel. Plutonium naturally emits neutrons. They are absorbed by the U-238, which through a series of decays forms more plutonium, which is used to resupply the fuel core—the step that is called "breeding." As the Department of Energy summed it up, breeder reactors "can stretch our nuclear fuel supplies while producing electricity."

Manhattan Project scientists first conceived of a breeder reactor in the early 1940s. They were particularly keen on the concept because at the time it was estimated that the entire world's supply of uranium could generate the electricity needed by the United States for less than two years. Hence, a breeder reactor that could vastly multiply the power of uranium seemed to be highly desirable. "The country which first develops a breeder reactor will have a great competitive advantage in atomic energy," Enrico Fermi predicted.

By 1948, General Electric, with funding from the AEC, had launched a full-fledged effort to build one. Before long, the breeder was being touted as the magical solution to the nation's—and indeed the world's—energy needs. Breeders, advocates argued, would produce such abundant energy as to make possible the air-conditioning of Africa and the heating of the subarctic. Industry's wild-eyed passion for the concept was summed up in a comment Carl Walske, president of the Atomic Industrial Forum, made some years later: "Telling a utility executive that he can't have a breeder is like telling a man he can't have grandchildren."

Conceptually, the breeder is extremely simple. Like communism, though, the idea worked best on paper, and the few attempts to build a breeder have resulted in some of the scariest episodes of the nuclear era.

Breeders are far more dangerous than conventional reactors and hence make engineering more difficult and expensive. A conventional reactor can suffer a meltdown and spread radioactive material, but it can't blow up. Fission can only lead to an explosion—as with a nuclear bomb—if there is enough uranium or plutonium in one place *and* if the fuel is detonated with a massive implosion of additional fissionable material that produces a wave of neutrons. In a standard reactor, relatively small quantities of uranium are contained in separate fuel rods. A breeder reactor not only uses more highly enriched fuel but in the process of breeding produces huge quantities of neutrons. Hence, a core meltdown at a breeder can produce a nuclear explosion capable of blowing up a plant's containment building.

And there's more. Water is a perfect moderator for a standard reactor but not for a breeder. That's because water slows neutrons down so much that it reduces the chance of a successful bombardment of the uranium-238 blanket and production of new plutonium fuel. Therefore, breeders are cooled with liquid sodium, which transfers heat as efficiently as water but doesn't stifle neutron potency. Because sodium is itself so explosive, the welds and fittings in breeder reactors must be of much higher quality than those in light-water reactors. In addition, sodium's volatility means that the fueling of breeders must be done entirely by remote control.

Furthermore, by the early 1950s, the navy's Admiral Hyman Rickover was running a program that developed more efficient conventional light-water reactors, and the government had uncovered huge deposits of uranium in the Rocky Mountain region and the southwest. All this made the breeder far less attractive as a commercial reactor.

Breeder backers also disregarded a whole host of problems, including the new technology's potential to abet the proliferation of nuclear weapons. A conventional thousand-megawatt reactor produces plutonium as a by-product—enough to make about twenty bombs per year. A breeder of the same size produces enough plutonium to make one hundred bombs.

Breeder projects have been plagued by the same

difficulties—cost overruns, engineering screwups, and technological failures—that beset conventional nuclear-power plants. William Lanouette provided the best account of the breeder's troubled history in a 1983 article in *The Atlantic*. As he recounted, the AEC approved construction of the world's first experimental breeder in November 1947, estimating that it would take eighteen months to build the facility. It was four years before the plant, called EBR-1 and situated in Idaho Falls, Idaho, was finally able to sustain a chain reaction. A few months later, EBR-1's core, shaped like a football and weighing 114 pounds, produced enough heat to power the plant's turbines—thereby marking the planet's first generation of electricity from nuclear power. More good news came in June 1953, when chemists at EBR-1 separated a few milligrams of plutonium from the plant's uranium blanket. It was a tiny amount, but it proved that breeding was technically possible.

It was not long, however, before disaster struck. On November 29, 1955, the crew of EBR-1 was running a series of tests at the request of the AEC to see if the core could run with less of the unstable sodium coolant. It couldn't, and the ensuing chain reaction quickly threatened to escape control. A supervisor frantically ordered the operator to "scram" the reactor, but the employee was chatting happily with his wife on the control-room telephone and didn't hear the command. The supervisor leaped over a table in the control room and hit the scram button, but he was too late. The core overheated, and more than half the fuel melted, blowing radioactive gas out of the building. No one inside was harmed, but the facility was contaminated so badly that it had to be permanently shut down.

In 1958, the government built a new test reactor, EBR-2, at the Idaho site. In some ways, EBR-2 was a major success. It ran for nearly forty years—which has turned out to be the normal operating life for nuclear-power plants—generated electricity, and was used to test different types of breeder cores. But the reactor never bred a single milligram of new plutonium fuel and was shut down in 1994.

The project that would see the mother of all atomic screwups was launched in 1955, when a consortium headed by twenty-five electric utilities formed Atomic Power Development Associates. The new group sent the AEC a proposal to build a hundred-megawatt breeder near Detroit—just about forty miles south of the suburbs where David was raised. The breeder would sell steam for electricity to Detroit Edison, the consortium's lead partner, and plutonium to the federal government for bomb building.

The Enrico Fermi Atomic Power Plant was a fiasco from the very outset. The project was slowed by technical problems for years, and in 1962 the first of several sodium explosions at the plant, still under construction, ripped out part of the reactor's coolant system. "Sodium-water 'reactions,' as the plant staff called them, became so frequent that the principal steam-generator pipes were named after the sound they made: Ker-Pow #1, Ker-Pow #2, and Ker-Pow #3," Lanouette wrote.

In 1963, the Fermi breeder finally went "critical"—to use the nuclear industry's odd term for a plant whose core is capable of producing fission. That was four years behind schedule, and more safety and repair problems kept it from generating electricity for the commercial market until 1966. Within three

months, the Fermi plant self-destructed. The cause was faulty
safety devices called conical flow guides, which were supposed to
improve the flow of sodium coolant to the core. When several of
the flow guides malfunctioned, the sodium flow was cut off,
causing part of the plant's core to melt and the reactor to be
scrammed. Fortunately, the meltdown was contained, and no
one was harmed, but it took months before operators figured out
what caused the core to melt, how much damage had been done,
and how to deal with the radioactive debris. Meanwhile, the
plant was shut down.

The AEC's own figures showed that if the Fermi plant had
experienced a full-scale meltdown, as many as 3,400 people
could have died and an area roughly the size of Pennsylvania
contaminated. And John Fuller, author of *We Almost Lost Detroit*,
which chronicles the Fermi incident, said the AEC seriously
underestimated the worst-case scenario. He cited a University of
Michigan study that placed potential casualties at 133,000. "The
task of evacuating Detroit would be flatly impossible because the
automobile city had put all of its faith in motor transport," Fuller
wrote. "It had never built a subway or elevated transit system.
During commuting hours, traffic was impossible. . . .
Coordination with the Canadian Civil Defense in Windsor would
be futile. The tunnels and bridges would be packed to capacity."

As the Fermi accident unfolded, the plant management
called the local sheriff's office and notified state-police
headquarters about the "problem" at the facility. Yet even as they
mulled over the necessity of evacuating the area, plant and
government authorities were careful to keep the press and the
public in the dark. On the day of the debacle, no one notified

local newspapers or TV stations, so no word of the accident reached the public. The next morning, the Fermi plant's top nuclear engineer, Walter McCarthy, released to a small area newspaper a communiqué that Fuller described as being "couched in terms reminiscent of a wartime battle report." There was no mention at all of the plant's fuel having melted; the only unusual thing noted was that "the radioactivity level of the [plant's] argon gas was observed to rise substantially." (It was almost two decades before reporters and independent investigators uncovered the full story.)

This spin marked a sharp contrast to the official posture when it came to advertising the threat of a Soviet nuclear attack on the United States. The government stoked fears about that possibility, in part to build support for its own nuclear-bomb-building aspirations. When it came to the threat of a nuclear-power-plant disaster, a far likelier prospect than a Soviet nuclear strike, government and industry preferred public complacency—or, better yet, ignorance.

At the time of the accident, the Fermi plant had eaten up $120 million while producing just fifty-two hours' worth of electricity. Instead of producing power that was too cheap to meter, the plant's electricity was coming at a cost of more than $2 million dollars per hour. And the Fermi plant did not produce any plutonium whatsoever—in other words, it didn't breed new fuel—because, as in every other breeder-reactor experiment, an elegant and simple theory proved unworkable in practice.

In 1970, the plant was restarted but ran only intermittently. By 1972, the AEC had conceded that the Fermi reactor was a lost cause. The plant was shut down after having run at full power for only about one hundred days over nine years. It still stands as one of the most spectacular commercial flops of the nuclear era.

Government and industry struggled onward in their efforts to get breeders up and running, always against the backdrop of nuclear failures around the world. In late 1957, an explosion ripped through a military-run plutonium-production facility in the Russian Urals. The accident killed hundreds of people and contaminated a huge swath of land. According to the book *Nukespeak*, thirty towns were evacuated, and their names were subsequently erased from the map. The CIA knew about the incident almost as soon as it occurred, but since disclosure of the catastrophe would have been hugely embarrassing to Soviet and American nuclear planners, both governments tried to keep word of it from reaching their respective publics. The story came out only in 1976, when Zhores Medvedev, a Russian biologist and dissident, published an article on it in the British journal *New Scientist*. The U.S. and Russian governments ignored the story, while Sir John Hill, head of the British Atomic Energy Authority, denounced Medvedev for spreading "rubbish" and "pure science fiction."

The same year as the Urals disaster, another nuclear nightmare was brewing in England, and the government, like its Soviet counterpart, hushed up the whole affair for decades. The accident took place at the Windscale plutonium-production reactor on England's northwest coast, near Cumbria. Windscale was part of the British government's secret effort to build an atomic bomb, as the United States, seeking to prevent other world powers from obtaining nuclear weapons, had denied London access to the relevant technology. Due to faulty instrumentation, a technician believed that the reactor was cooling down too much and gave it an extra shot of heat. This quickly produced a fire that partly consumed eleven tons of uranium. A huge radioactive cloud was sucked through a chimney and spread throughout the countryside and as far away as Denmark.

To keep the extent of the disaster hidden, the government decided not to evacuate people living near the plant, even though they were exposed within hours to a dose of radiation ten times higher than the allowable lifetime amount. The fire raged for sixteen hours, and scientists managed to control it only after they flooded the reactor with cooling water. It took until 1983 for the British government to reveal that thirty-nine leukemia deaths could be traced to the Windscale accident. After finally confessing the incident, the government tried to wipe the slate clean by changing the plant's name to Sellafield. The plant was still being disassembled in 2003, and the molten uranium that had burned yet emitted a gentle heat.

In 1961, a nuclear accident occurred that couldn't be covered up, when an explosion in the control room of the SL-1 experimental reactor near Idaho Falls, Idaho, killed all three workers on hand. The cause of the explosion was never determined. One theory is that it was simple human error on the part of a worker who pulled a control rod out too far, thereby setting off an uncontrolled chain reaction. The other supposition, more baroque, is that one of the employees deliberately pulled a control rod out of the reactor because he believed a coworker was having an affair with his wife. (A subsequent AEC report delicately described this possibility as "malperformance motivated by emotional stress.") The blast destroyed the reactor and spread radioactive material across fifty acres of sagebrush.

Six years later, winds spread radioactive particles from waste stored at Russia's Chelyabinsk-65 complex, and nine thousand people were temporarily evacuated. Back in the United States, a 1971 accident at Northern States Power's reactor in Monticello, Minnesota—which took place when a water-storage space overflowed—dumped fifty thousand gallons of radioactive wastewater into the Mississippi River.

Though nuclear accidents were relatively infrequent (if underreported) and produced few direct casualties, the potential scope of such catastrophes began to undermine public support. In Washington, D.C., however, nuclear power's image remained unscathed. This was largely due to the unbridled enthusiasm of Glenn Seaborg, who had become head of the AEC in 1961 under John F. Kennedy. Seaborg was one of the most brilliant scientists of the nuclear age, but he had an almost maniacal devotion to atomic power and was remarkably blind to any negative consequences that it might produce. Seaborg saw the atom as a construction tool, not just a source of power. In one essay, he pondered a nuclear strike on the Strait of Gibraltar as a means of irrigating the Sahara, though he admitted that the advantages "would have to be weighed against the loss of Venice and other sea-level cities."

As chairman of the AEC, Seaborg laid out his vision of a "plutonium economy" of the future, in which breeders would help ensure that human civilization was not "just a brief fossil-fuel flicker in the long cosmic night." Seaborg's influence was clear in the Nixon administration's national energy policy, announced in 1971. In addition to proposing hefty subsidies to the nuclear industry and the elimination of public input into licensing decisions, the president's plan promised to end U.S. dependence on foreign oil by 1980. To ensure this, nuclear plants would be called on to meet half of the nation's energy needs by the dawn of the twenty-first century, thereby requiring the construction of a network of as many as one thousand reactors. Energy independence also meant full speed ahead for breeder reactors, which Nixon called "our best hope today for meeting the nation's growing demand for economical, clean energy."

Out of this policy came the Clinch River Breeder Reactor in Tennessee, which was another joint collaboration between the AEC and private industry. Despite a record steeped in failure, scientists and breeder builders maintained their optimism, believing that technological advances would ultimately provide them with answers. Hence, they gaily conceived of building bigger cores that would produce more energy. Clinch River was to have a core weighing 22,000 pounds, dwarfing the 114-pounder that had melted at EBR-1.

Spearheaded by Al Gore, then a member of the House of Representatives from Tennessee, the project received lavish funding from Congress. But even as the checks were going out, estimates of the plant's cost began to increase dramatically from the AEC's original 1971 budget of $500 million. The private companies, experienced with reactors coming in over budget, convinced Congress to require the federal government to pay for all cost overruns. The utilities would never have to put up more than their original stake of $250 million, thereby ensuring that they had no incentive to keep a lid on costs.

By 1974, estimates for the building of the Clinch River plant had reached $1.7 billion. A few years later, the Department of Energy's estimate rose to $3.6 billion, while the General Accounting Office, an agency more independent of industry, put the figure at $8.5 billion. In 1983, when it was estimated that completion costs would deplete much of the federal budget for energy research and development, Congress finally killed the Clinch River program. By then, the government had shelled out almost three times more than the original estimate to build the entire plant. Almost all of that money had gone for research and planning; construction never got much beyond laying a foundation.

Even after the Clinch River disaster, breeder backers refused to concede defeat. Advocates at EBR-2 in Idaho Falls and at another research station, the Fast Flux Test Facility at the Department of Energy's Hanford site in Washington State, continued to press their cause and urged Congress (unsuccessfully) to appropriate more funds for further research. The Department of Energy sponsored a competition between General Electric, Westinghouse Electric, and Rockwell International to see which could develop a new model for breeder development. "Despite years of setbacks and daunting economics, [breeder] technology still has determined supporters who declare its opponents shortsighted," *Nucleonics Week,* a trade publication, declared in 1988. "The dream of nearly limitless energy continues."

Beyond Idaho Falls and Hanford, there weren't many places where the breeder dream still burned brightly. But one of the few was Commerce Township, Michigan, where David Hahn would soon stage the final act of his nuclear research.

CHAPTER 7

From Theory to Practice:
How the Potting Shed Came to Glow

Getting useful energy from atoms took several more years.
People could not build reactors in their homes.

—THE BOY SCOUTS ATOMIC-ENERGY
MERIT-BADGE PAMPHLET, 1983.

s David went about his business of collecting radioactive
elements and conducting haphazard experiments, he
continued to research every aspect of atomic history, from the
uses of the elements to the building of the atomic bomb.
Inevitably, his attention turned to the subject of nuclear
reactors, where the elements that so intrigued him were

combined to generate the most basic yet wondrous end product of the atomic age: the chain reaction.

For help in understanding the principles of the chain reaction, David turned frequently to the Department of Energy for information, sometimes as the intrepid Professor Hahn and sometimes as a shy, polite Chippewa Valley student doing supplemental research. One of the items he received in reply was a copy of *Atoms to Electricity*. In addition to boasting about the still untapped potential of nuclear power, the booklet made breeder reactors out to be an "old" and essentially proven technology, stating that scientists and engineers had been working on their design for more than three decades. Furthermore, it asserted, several European nations, not to mention Japan and the Soviet Union, were quickly moving ahead with their own breeder programs. If the United States didn't follow suit, went the obvious if unspoken conclusion, those nations and perhaps others would soon leave the United States in the economic dust.

From the *Golden Book*, his scouting years, and the Department of Energy's propaganda handouts, David learned only of the triumphs of atomic engineering. He knew little of the multiple setbacks suffered by the nuclear industry, nor did he know that the breeder reactor was, for all practical purposes, a pipe dream. He knew virtually nothing about the history of the Enrico Fermi Atomic Power Plant, even though that colossal failure had unfolded practically in his backyard. David had never even heard of the Clinch River Breeder Reactor. And though he'd done some minimal reading about the accidents at Three Mile Island and Chernobyl, his nuclear education had left him numbed to the consequences of those events.

Hence, despite coming of age at a time in which all the assorted screwups and accidents of the atomic age had generated a powerful antinuclear movement, David remained comfortably cocooned within the confident optimism of the 1950s and 1960s. His passion for the atom was fueled by the conservative family and community environment in which he was raised, as well as a natural aversion to intellectual challenge (which does not bode well for a career in science). "I tried not to read anything that would disappoint me or make me negative," he said candidly of his research strategy. "If I knew it had a critical perspective, I wouldn't even pick it up."

Now, as he read about nuclear-power plants and pondered the future, David had reached a crossroads. He might well have chosen to bring his radioactive research to a close and turned his attention to new scientific challenges. After all, he'd collected most of the elements on the periodic table, including a number of radioactive ones. What more could he hope to achieve? The answer presented itself as David was aimlessly browsing through one of his father's old college textbooks, which contained a chapter on the history and virtues of the breeder reactor and a discussion of how such reactors worked.

By now, he had some familiarity with breeders and was intrigued by the idea of a perpetual-energy machine. However, sources such as *Atoms to Electricity* discussed breeders in infuriatingly abstract (albeit glowing) terms. Now, courtesy of the textbook, he had a schematic drawing of a real model, laid out in a checkerboard pattern, with alternating cubes of radioactive fuel and moderating elements to regulate the chain reaction.

The insights meshed marvelously with research he'd done earlier, when searching out sources of thorium for his collection. A scientific encyclopedia he'd consulted at the time had pointed out that nonfissionable thorium-232 becomes thorium-233 when it absorbs a neutron in its nucleus. Thorium-233 emits a beta particle during radioactive decay and becomes uranium-233— a man-made fissionable element that can be used in place of plutonium as fuel in a breeder reactor. Because of its natural abundance, thorium "represents a tremendous potential source of nuclear energy," David had read in the encyclopedia.

Putting one and one together, David figured that he now could find the components needed to assemble a model breeder reactor—and this time it wouldn't have a juice-can core but a real one, pulsing with radioactive fervor. He'd already found thorium-coated lantern mantles, which could be transformed into uranium-233—the perfect fuel for a breeder! He'd need neutrons to bombard the thorium-232, but that wouldn't be hard: Neutrons can be produced by combining a decent alpha-emitting element—such as americium from smoke detectors— with aluminum. If he could accumulate enough uranium-233 to sustain a chain reaction—about thirty pounds is needed for a critical pile—he just might be able to pull off an incredible feat. "I was pretty amazed when I saw the checkerboard drawing," David remembered with excitement. "I thought, 'Holy smokes, I can go with this!' I already had a lot of the materials I needed, and it seemed that it wouldn't be that hard to get it to work."

Of course, there were a few other kinks that would need to be worked out. He'd have to isolate and purify the thorium-232 from lantern mantles, for example, and find a source of uranium for the "blanket" needed to breed new fuel, but these were difficulties that could surely be overcome with a little muscle and brainpower. Anyway, there was no point in being negative; he'd tackle these problems when they arose. Then he'd assemble his materials into a workable model and join the Curies, Fermi, and Otto Hahn among the ranks of radioactive superstars.

David was aware—at an intellectual level, anyway—that thirty pounds of uranium-233 was far more than he could realistically hope to produce. But he figured that if he could put together a checkerboard model breeder with enough radioactive raw materials, he'd at least be able to transmute the atoms of his various isotopes of thorium and uranium into wholly new elements. For David, transforming an element—altering its natural, God-given chemical makeup!—approached a sacred act. "I'd imagine an atom of uranium that had been sitting around, in all practicality, for eternity," he said. "Then a stray neutron enters it, and a new birth takes place. When one element is transformed into another, the universe is slowly recycling itself."

So at the ripe old age of sixteen and a half, David set out to succeed where armies of trained scientists with infinite resources had failed. He may have been naïve in thinking that he could build a breeder reactor in his backyard, but he was no more starry-eyed and arrogant than the scientists, businessmen, and government officials who came before him, with their grand plans to use nuclear bombs to irrigate the Sahara, Venice be damned.

Young people who achieve spectacular success in science generally do so at school, where they work closely with teachers, have structured schedules for supervised research, and frequently work in teams. In other words, they are part of a community that pushes them to achieve amazing things and rewards them when they do. Consider past winners of the Siemens Westinghouse Competition in math, science, and technology for high school students. Based on their résumés, they share one trait with David: They're classic science dweebs. They join the math team and Model United Nations, compete in the Physics Olympiad and Academic Quiz Bowl, play the violin and chess, and for fun belong to the Mars Society and Future Business Leaders of America.

At that point, however, the similarities end. For starters, the Westinghouse champs pursue projects that are far more practical than David's and in some cases make truly valuable contributions to society. In 1999, Lisa Harris of New York City won for developing a procedure to identify four common carrier genes of cystic fibrosis. In 2000, Heidi Hsieh, a Smithtown, New York, resident, developed "new structures that may have possible applications as filters for larger macromolecules such as DNA." The 1999 team winners, Daniar Hussain of Johnstown, Pennsylvania, and Steven Malliaris of Winnetka, Illinois, discovered improved methods of data storage and retrieval. Their pragmatism was surely grounded in the fact that they worked closely with professors, mentors, and like-minded high school colleagues.

With similar guidance, David's research might have taken a far different and more productive course. Of course, his very makeup made it difficult for him to work in a group or even to understand how a group operates, but even if David had been so inclined, almost no one showed any real interest in what he was doing or was clever enough to recognize his talents. The one exception was Jim Hungerford, the father of David's friends Andy and Jeffrey, who was active with Troop 371. Jim spotted David's quirky genius and pushed him to consider a career in science, but he didn't see David often, and so his prodding didn't lead anywhere. "I tried to talk to my teachers, but for most of them it was just a job," David said. "They saw my stuff as a hobby, like stamp collecting."

Meanwhile, Ken Hahn viewed David's scientific interest as an unhealthy obsession, something that scouting or other diversions could cure. His mother remained as supportive as ever, but she was in no position to provide her son with any direction. Patty's drinking had gotten heavier over the years, and the effects were made worse by the medications she took for schizophrenia and depression. She'd usually make it through the day when David was around, but when the sun went down she and Michael would drink beers or shots of hard liquor.

Hence, David went about his work in complete isolation, following his own strange impulses in developing his typically idiosyncratic strategies and paths. He improvised, frequently brilliantly but—in part because he had no peer review—almost as often obtusely. Every decision was his and his alone. Years before, David had founded the Big D Lawn Mowing Service; now he would establish the Big D Nuclear Power Plant.

Before embarking on his breeder-reactor project, David gave a makeover to his potting-shed laboratory, the primary site of what was to be a yearlong odyssey. He applied a fresh coat of white paint, put a colorful chart of the periodic table of the elements on the wall, and installed a patch of brown carpeting on the floor. He also upgraded his lab equipment with a few Pyrex beakers and vessels that he bought from a chemical-supply shop.

The lab looked spanking new, though it still provided its director with virtually no protection. At a real laboratory where radioactive research is performed, the air is ventilated, filtered, and monitored constantly for leaks. Staff members are provided full protective gear, such as face masks, ventilators, and glove boxes—sealed metal chambers with attached lead-lined gloves that allow workers to keep from directly handling deadly materials. David's chief precaution was still his lead-lined civil-defense suit, which by now was ripped and chemically damaged. His sole safety upgrade now was the installation of a thin, malleable sheet of lead he found in a friend's garage as a crude shield. He placed it around a section of tabletop that served as his workstation in hopes that the lead would block emissions from the radioactive elements he hoped to acquire and manipulate.

David was anxious to get started, but his first step, a prudent one by his standards, was to seek out additional technical information and advice. Once again, he sent away for help. No single source was more obliging than the Department of Energy, where Professor Hahn managed to engage the director of isotope production and distribution, Donald Erb, in a scientific discussion by mail.

"Does aluminum produce neutrons as well as beryllium with an alpha reaction?" David asked Erb in one letter. The question seemed innocent enough, but a satisfactory answer was vital to David's unfolding plans: He planned to build a "gun" that would fire neutrons at thorium-232, thereby transforming it into uranium-233, his breeder fuel. Erb, oblivious to the fact that Professor Hahn was in fact a Boy Scout hell-bent on building a model reactor, delivered the goods. "Nothing produces neutrons from alpha reactions as well as beryllium," he replied by mail, specifying that it could be up to 250 times more efficient as a neutron producer than aluminum.

David had planned to combine aluminum with an alpha-emitting element to create neutrons, but as he excitedly read Erb's letter in his basement bedroom, he began contemplating how he might obtain beryllium, a light, gray metal that is six times stronger than steel and used to build missiles, rockets, and communication satellites. Erb imparted another useful piece of information to David when he informed him that experts at the government's Los Alamos laboratory in New Mexico had determined that radium, like americium, is an energetic "alpha emitter"—the second component David needed to generate neutrons. Once again, David altered his plans. He'd always had a soft spot for radium—he was captivated by its glow, and it was, after all, the prize discovery of Marie Curie—and so he added it to his new shopping list.

Erb also provided a list of isotopes that can sustain a chain reaction, explained the relative ease with which different atoms would fission, and threw in tips on purifying thorium from thorium dioxide. This latter subject was of critical importance because gas-lantern mantles are coated with thorium dioxide, a less radioactive form that would need to be purified before it could be transformed via neutron barrage into uranium-233 breeder fuel.

When David asked Erb about the risks posed by such radioactive goods, his pen pal—assuming that a combination of the law, insufficient technical expertise, and common sense would keep any individual from acquiring such items—informed him that the real dangers were slight since possession of radioactive materials "in quantities and forms sufficient to pose any hazard is subject to Nuclear Regulatory Commission (or equivalent) licensing." Of course, the NRC would not license a backyard shed, but it would shut it down if it was discovered. As he digested Erb's missive, David became increasingly determined to be tight-lipped about his activities, not just with family members but with his few friends as well. (Erb has retired from the Energy Department and could not be reached for comment about his correspondence with David.)

On occasion, David still discussed his radioactive aspirations and progress with Heather. Since she wasn't all that interested in the subject and had no idea what he was talking about, she missed quite a few red flags.

The only other person David regularly confided in about his breeder project was Jim Miller, his friend and classmate at Chippewa Valley. Like David, Jim was a loner who was failing to live up to his academic promise. Jim was also a math whiz and helped David with his algebra homework. David did tell his friend about his plans and, aware of the complex challenges he would face in the months ahead, asked him if he'd like to help out. Jim had enough common sense to turn down the offer, but he was happy to serve as a sounding board and even made David a helpful three-dimensional drawing of a model breeder reactor.

Undeterred, David moved to step one of his project: building a working neutron gun to transform the thorium and uranium that he planned to acquire. When in 1830 Samuel Colt was first struck with the inspiration for his soon-to-be-famous revolver, he sketched out a crude drawing and carved a model out of pine. David did more or less the same thing upon concocting a plan to build a neutron gun. His design, which he quickly implemented, called for a block of lead to serve as the gun's frame. Lead is soft and malleable, and David easily hollowed out the block with a chisel, then smoothed out the pit with the narrow head of a power sanding tool. Into this chamber would go an alpha-emitting element and a quantity of beryllium or aluminum, both of which absorb alpha rays and in the process kick out neutrons. Last, David pricked a tiny hole through the lead block with a needle so the submicroscopic neutrons could stream out to their target.

David's gun was ready, but he couldn't produce his neutron "bullets" without first obtaining beryllium or aluminum. He knew from Erb that the former would produce more neutrons than the latter, but he couldn't round up any beryllium, despite numerous calls to chemical suppliers. David didn't then know it, but beryllium can be used to build atomic bombs—when wrapped around a core, it augments a chain reaction—and so can be bought only from a few suppliers regulated by the Department of Energy. In its solid form, beryllium poses only a modest health risk, but it is carcinogenic if inhaled. Eight workers who helped assemble the earliest atomic bombs died of cancer caused by beryllium dust. Hence, the Occupational Safety and Health Administration and the EPA also control its circulation. But whatever the causes for his failure, David knew when to cut his losses. He settled, at least temporarily, for common aluminum strips bought from a chemical supplier.

He also had to compromise on the issue of an alpha emitter. His first choice was radium, which he planned to recover and isolate from the painted faces of antique luminescent clocks. With Geiger counter in hand, he began visiting junkyards and antiques stores and rummaging through attics and basements in the homes of friends and relatives. After hunting for a few weeks, he'd collected several dozen old clocks. David spent an entire sunny weekend sitting beside the potting shed with the pile of clocks on one side and a pitcher of lemonade on the other (covered with a towel to keep it from being contaminated). With a screwdriver, he chipped away the paint from the clock dials and collected the radium flakes in pill vials. It was a highly dangerous process—radium dust can be ingested through the mouth and nose—and a terribly inefficient one. As nightfall came that Sunday, David had only a small pile of radium chips to show for his trouble.

With radium seemingly a lost cause, David considered using polonium as an alpha source. But when he tested the silver strips he had stored from electrostatic film brushes, the Geiger counter picked up very weak alpha emissions, probably because the polonium-210 in the strips has a half-life of only 138 days. He discarded polonium, too, as an alpha source for his gun.

Americium-241—the isotope found in smoke detectors—has a more resilient half-life of 453.1 years. David still had a few smoke detectors from his raid on the Lost Lake cabins, but those wouldn't be enough for his current purposes. The single americium chip found in a smoke detector emits a tiny amount of alpha particles; David calculated that he'd need at least one hundred detectors to accumulate enough americium. Buying them at retail prices of up to fifty dollars a pop was out of the question—David's minimum-wage jobs hardly allowed for such extravagances—and frustration was starting to set in. Maybe building a nuclear reactor wasn't as simple as it seemed.

Then Jeffrey Hungerford saw a newspaper advertisement announcing that a local company was selling off at bargain-basement rates loads of water-damaged merchandise, including smoke detectors. David drove to the store and purchased every one in stock, which came to a few dozen.

At almost the same time, David read somewhere (possibly in *Popular Mechanics*, to which he subscribed) that companies such as First Alert and Captain Kelly sell off at a huge discount defective and expired smoke detectors. Claiming that he needed a large number for a school project, David convinced one of the firms to sell him about one hundred for a dollar apiece. Soon, the post office was delivering boxes and boxes of smoke alarms to the Hahn household. When Ken asked David about the mysterious packages, he got (and swallowed) the usual feeble story about a scouting project in the works.

David used pliers to remove the americium chips, which he stored in a pill vial. Then he bided his time. One afternoon, when Ken and Kathy were at work and his stepsister Kristina was at a friend's house, David pulled a blowtorch from its storage spot in the garage. Wearing only a paper nurse's mask and gloves for protection, he poured the americium chips into a pan that he placed on the patio table. He fired up the blowtorch and spent two hours welding the chips together. When he was done, he was the proud owner of a small, misshapen, silver-black marble of americium.

With americium and aluminum, David had the materials he needed to make neutrons. But how could he test the gun? Neutrons don't make noise when they're "fired," and since they have no charge a Geiger counter cannot measure them. At this point, David recalled something that the Joliot-Curies first discovered and reported to the French Academy of Sciences in 1932. When paraffin is hit by neutrons, it throws off protons; protons emit a charge and therefore can be detected by a Geiger counter.

David bought a block of paraffin from the drugstore and drove to Michael and Patty's house, armed with his neutron gun, americium ball, and aluminum strips. Relieved to find that no one was home, he pulled his lead-lined suit out of the closet and put it on over his jeans. In the potting shed, David loaded the neutron gun's chamber with americium and aluminum and placed the block of paraffin on the table. He put the gun beside it, with the hole in its side aimed at the paraffin. The absolute silence in the shed was broken by the rapid clicking of David's Geiger counter, registering a proton stream reflecting off the paraffin block.

David was overjoyed yet unbelieving of his own success. He dashed from the shed into the house and raced up the stairs to the bedroom, where he kept a copy of *Modern Chemistry*. Flipping through the pages, he came to the section that described the reaction caused by neutrons hitting paraffin. As his eyes raced across the text and he saw that he had indeed succeeded, he felt an almost overwhelming sense of pride and happiness.

David's neutron gun, crude but functional, was ready for action. He put the lead block on the shelf of the potting shed, next to the americium and aluminum, which he stored in separate glass containers. With the first step of his project complete, it was time to turn to step two: procuring fissionable elements to use as radioactive fuel.

Decades earlier, northern Michigan had been the site of freelance and corporate mining. David read up about the region's geologic and mineral history and even found an old prospector whom he interviewed. Not long after successfully building his neutron gun, David, who'd recently turned seventeen, spent the whole of several Saturdays in his Pontiac, scouring hundreds of miles of upper Michigan in a search for "hot rocks" containing uranium ore. Accompanied only by the sound of pop music blaring through his sound system—and his trusty Geiger counter, mounted as always on the dashboard—he drove through state parks and parked near trailheads on the sylvan terrain. Detaching the Geiger counter from the dashboard, David would set off with high hopes, only to return to the car empty-handed. When it was time to drive back to Detroit, all he'd found was a quarter of a trunk load of pitchblende on the shores of Lake Huron.

At the potting shed the next day, David began the arduous, miserable task of pulverizing his pitchblende with a hammer. At least, he thought, he was truly following in the footsteps of the Curies, who'd done precisely the same thing nearly a century earlier with their Bohemian treasure trove. It was small consolation, especially when the sample of the smashed ore he tested with his Geiger counter appeared dead.

David decided to pursue a more practical approach. When the NRC had sent him its list of sources and pricing information for radioactive materials, David had noted that two firms in the Czech Republic sold small uranium samples, one to commercial and university buyers and another to museums and specimen collectors. He wrote the first company in the guise of a professor seeking materials for simple laboratory experiments. He asked for information and samples of uranium-bearing ores. No subterfuge was needed in the case of the second. David simply enclosed a money order for $140, which covered the cost of a marble-size sample of uraninite (a radioactive mineral) and a brick of pitchblende.

While David waited anxiously for news from the other side of the Atlantic, he carried on with research on other fronts. By now, his life beyond science consisted of going on dates with Heather, pursuing Eagle Scout standing, and daydreaming his way through school while doing just enough to keep from failing. David had earned the last of his merit badges but still needed to complete the final requirement, a service project that benefited the community. David designed a patio and pathway for the public library and convinced a local construction company to donate the half ton of concrete and crushed rock needed to build it.

David's romantic life was not proceeding as smoothly. Heather, his one remaining nonradioactive passion, had begun to do something her mother had proposed a few years earlier:

ponder her future with David. It would have been a remarkable woman indeed who could have endured his particular brand of courtship, especially someone of Heather's temperament. She was, like David, a dreamer, but her aspirations were tempered by a pragmatic sensibility. Heather wanted to become a successful businesswoman, make her first million dollars, and buy a house with an indoor swimming pool. It was a lofty goal, but Heather had a sensible plan: First she'd obtain her undergraduate degree, and then she'd get a master's in business administration. After that, it was the executive track, hopefully for a firm with overseas operations, since Heather wanted to travel.

In David's case, pragmatism was in short supply. He wanted to become a famous scientist whose name would, he told Heather, "go down in history." But David's plan and vision didn't extend far beyond conducting research in his potting shed until something earth-shattering took place. While many of his peers at Chippewa Valley were thinking about where they would like to go to college and worrying about SATs, David hadn't given the slightest thought to future studies.

Even between high school sweethearts, such fundamental differences inevitably became evident. Furthermore, David's research made it almost impossible to have a normal romance. He worked on his projects every day, even weekends. When they did get together, Heather would try to start a "normal" conversation with him, but David would soon take a fork in the road that led back to his research interests. If things didn't change soon, Heather thought, the relationship was finished. Even oblivious David was picking up signs of trouble—when he brought up his scientific activities, Heather's tight smile made clear she didn't want to hear about it—though he didn't understand how close he was to losing his girlfriend. In any event, he was far too obsessed to stop his research now that he'd come so far.

Several weeks after writing to the Czech Republic, David received his replies. Professor Hahn received a polite letter and a few samples of a black ore—either pitchblende or uranium dioxide, both of which contain small amounts of uranium-235 and uranium-238. Specimen collector David Hahn was sent a slightly larger package containing the uraninite and pitchblende that he'd ordered.

At the first opportunity, David was back in the potting shed. He took a hammer to his new samples and was cheered when his Geiger counter showed that the crushed ore was livelier than the tepid pitchblende from up north. During the prior weeks, in a chemistry textbook at the local library, he'd found a recipe for isolating uranium isotopes from ore that called for a mixture of nitric acid and sulfuric acid. He purchased the latter, a potent but common chemical used to make everything from gasoline to detergents. For nitric acid, he used a saltpeter-and-sodium-bisulfate-based formula he'd first manufactured years earlier when brewing nitroglycerin.

He then mixed the nitric acid and sulfuric acid with the powdered ore and boiled it, ending up with something that "looked like a dirty milk shake." Next, he poured the milk shake through a filter—he'd neglected to buy a proper one from a chemical-supply shop, so, necessity being the mother of invention, he used a coffee filter from the kitchen—hoping that the uranium would pass through while the rest of the sludge would be trapped behind. But David miscalculated uranium's solubility, and whatever amount was present was trapped in the filter, making it difficult to purify further.

What David was trying to do, in a simplistic way, was to make yellowcake, a product central to the industrial production of nuclear power. Yellowcake—the name comes from its yellow, powdery consistency—is produced by using acids to dissolve crushed uranium, and the result is then made fissionable through further enrichment and purification.

Fortunately, David's attempt to bake homemade yellowcake was doomed to fail. The procedure itself isn't terribly complicated, but separating U-235 and U-238 from uranium ore—using the magnetic-field or gaseous-diffusion technique discovered by Manhattan Project scientists and still in use today—calls for sophisticated and vastly expensive equipment. (The concept of an atomic bomb, which is basically an unregulated reactor, is fairly simple as well but also requires separating and enriching U-235. That's the primary reason that so few countries have been able to build one.)

Yet the attempt itself was an act of pure madness and reveals a lot about David's lack of concern for personal and community safety. For if he had ever actually secured a supply of the uranium-238 isotope and used a functioning neutron gun on it, he would have produced plutonium, the most deadly substance known to man. Where would he store it—in the refrigerator next to his mushroom shakes and chemical potions? This was another bridge to be crossed.

Frustrated with uranium, David turned his attention to

thorium. He'd already found traces of it in gas-lantern mantles, but now he required huge new quantities. David dropped by his original supplier, Ark Surplus, but the store didn't have any more lanterns. He called other hunting and camping surplus stores, to no avail. David considered concocting another of his bogus stories about a chemistry class project, but it seemed doubtful that Coleman or any other gas-lantern manufacturer would fall for a story, even from a genuine Boy Scout, about a student's urgent need for hundreds of thorium-dioxide-coated mantles.

Desperate times call for desperate measures, and thus David resorted to a method that he later described as being "not quite legal": He befriended a worker at a retail camping-equipment store and paid him to pilfer boxes of replacement mantles from a storage room. At an agreed-upon time, David drove his Pontiac to the back of the store, money exchanged hands, and David and the store employee loaded his trunk and backseat with boxes of brand-new mantles.

It took David an entire Saturday afternoon in the backyard at Golf Manor just to unwrap the mantles—there were hundreds of them by his estimate—from their packaging. Exuberant, David went "thorium crazy." He fired up a blowtorch in the potting shed, strapped on a gas mask, and set out to reduce the mantles to ash. He didn't stop until midnight, then went to sleep and started again early the next morning after gulping down a bowl of Frosted Flakes. By the end of the day Sunday, when Ken came to pick him up and take him back to Clinton Township, David had reduced all the mantles into a mountain of fine powder. He stored the ash in milk jugs, mason jars, assorted plastic bottles, and shoe boxes, some of which he left in the potting shed and some of which he hid in the back of a closet at Michael and Patty's house.

However, for best results, David would need to isolate and purify thorium from the ash before trying to irradiate it with the neutron gun. He had researched the matter while waiting for his uranium shipments to arrive and found a method in one of his dad's old chemistry books. There he learned that lithium, number 3 on the periodic table, is the lightest and most "reactive" of all metals. If certain steps are followed—and the textbook described them in detail—a lithium fragment will absorb bromides, chlorides, nitrates, dioxides, and other compounds of any element. At the same time, the lithium is transferred from its original fragment to the powder, strip, or other vessel that held the original element.

This might sound complicated, but for David's current purposes the consequences were as straightforward as they were significant. If he followed the steps correctly, he would start with lithium strips and thorium-dioxide ash and end with lithium ash and potent thorium strips that he could try to turn fissionable with his neutron gun.

Before trying out the technique on his precious ash, he resolved to conduct a trial run with a nonradioactive substance. If he could use "lithium replacement" to purify potassium from potassium nitrate, David would then employ the procedure on his mountain of ash.

David went to a Rite Aid a few blocks from his father's house and bought a container of potassium nitrate, which is also known as saltpeter and can be used as a curing salt for meats and to treat sensitive teeth. (It can also be used to produce gunpowder and explosives and is no longer available at drugstores.) Lithium was also easily available at retail outlets, but it was prohibitively expensive for David. Instead, with his cash reserves dwindling and unsure how much lithium he would ultimately require, he obtained the metal by shoplifting hundreds of dollars' worth of lithium batteries, which pack far more power than conventional batteries and are used in products such as laptop computers and cell phones. (The introduction of rechargeable lithium batteries in the 1980s is what made both of those items practical, as they dramatically reduced their size. Nokia's first mobile phone in 1982 included a bulky battery pack and weighed more than twenty pounds.)

In the shed, David carefully laid out the equipment on his lab table. He cut the batteries in half with a pair of wire cutters and used needle-nose pliers to pull out the shiny strips of lithium. Because lithium oxidizes—David could see the lithium darkening as soon as he cut through the batteries—he swiftly dropped the strips into a beaker filled with Crisco oil.

Then he placed about five lithium strips with a spoonful of potassium nitrate powder in a piece of aluminum foil shaped to the size of a small muffin. David turned a propane stove to high flame and heated up a pan of oil. When the shed smelled like a fast-food restaurant, he dropped the sealed foil ball into the pan and watched it skitter up and down in the oil for half an hour.

After allowing the foil ball to cool down, David opened it up. There was no indication that anything had changed; just as before, there was a small quantity of powder and a few silvery strips. But appearances can be deceiving. David had studied the reaction that occurs between water and all light metals; potassium, he knew, spins, ignites, and emits a velvety purple hue. David took a silver strip that had once contained lithium but should now contain potassium. He dropped it into a bowl of water on the tabletop, and the ensuing result was textbook perfect. And it was at this point, as he watched the strip spin and a purplish smoke arise from the bowl, that an ecstatic David truly began to expand his ambitions beyond the realm of simple irradiation; he believed that he might be able to create energy through nuclear fission.

Now it was time to use the lithium-replacement procedure on his thorium-dioxide ash, for which he donned his gas mask, lead suit, and latex gloves. Before putting the ash into the foil ball, he used his Geiger counter to test the strength of the thorium's radioactive emissions. He wrote down the results in one of the logbooks where he recorded all his breeder-related research observations. He repeated the experiment exactly as before and waited for the foil ball to cool.

His heart racing with excitement, David plucked from the foil one of the strips, in which lithium should now have been replaced with a cleaned-up form of thorium. He turned on the Geiger counter and moved it toward the strip. Eureka! The device was clicking with great intensity, and the readings of radioactivity were far greater than when the thorium had been contained in the ash. Indeed, it later was determined that David's method purified thorium to at least 9,000 times the level found in nature and 170 times the level that requires NRC licensing.

Over the next week, David spent almost every free moment in his lab purifying his thorium ash. Given the unventilated space he was working in, this was a highly reckless practice, particularly as he didn't always used his (tattered) lead suit and gas mask. He stopped only when he ran out of lithium batteries, and by then he had dozens of strips of thorium, which he stored in a glass jar.

Technically, David was now ready to turn his neutron gun on the thorium strips and remaining ash in hopes of turning the isotope into fissionable uranium-233. He still longed to substitute radium for americium in the gun but, having never found a sufficient source, was prepared to proceed without it.

Then, a stroke of luck. One day in the midst of his thorium frenzy, David was driving through Clinton Township on his way to pick up Heather when he noticed out the window a store called Gloria's Resale Boutique/Antique among a stretch of shops on a small street. He parked the car and took out the Geiger counter before heading into the store, which sold clothing, jewelry, furniture, rugs, lamps, china—and a large assortment of antique clocks.

When David strode into Gloria's, a few customers were milling about and a clerk was sitting behind a counter displaying antique jewelry and knickknacks. Everyone seemed bemused by David, and one person wanted to know why he was carrying a metal detector. His explained that the device was actually a Geiger counter, and the clerk and customers, by now intrigued, watched as he tested objects around the shop. The Geiger counter reacted to several clocks, but one in particular was jumping off the charts: a table clock from the 1920s or 1930s with a green-tinted dial.

The proprietor of the boutique, Gloria Genette, was at home that day when the clerk called to say that a polite young man was anxious to buy a clock but wondered if she'd come down in price. Gloria agreed to negotiate and, after a brief period of haggling, sold the clock to David for ten dollars. As soon as he got back to his car, David began taking apart the clock. Inside, he discovered a quarter-ounce vial of radium paint left behind decades earlier by a worker, either absentmindedly or perhaps as a courtesy so that the clock's owner could touch up the dial when it began to fade. The vial was sealed shut, as paint had dried around the cap, but it was almost full.

David was overjoyed. He dropped by the boutique later that night to leave a note for Gloria, telling her that if she received another "luminus [*sic*] clock" to contact him immediately. "I will pay any some [*sic*] of money to obtain one."

To concentrate the radium, David needed a small quantity of barium sulfate—a nonradioactive product but not available at the corner drugstore. Hospitals, though, stocked barium sulfate because it is radio-opaque and hence used to take X rays. Once taken inside the body, orally or via an enema, barium blocks X rays so that the resulting picture shows in white the shape of the organs that contain it.

Once again, David was quick to devise a plan. One day after school, he dropped by the X-ray ward at the local hospital where he'd fulfilled one of the requirements for the atomic-energy merit badge. Dressed in his scouting shorts and neckerchief in order to look his most saintly, he found a staff member who remembered him from his earlier visit. After hearing him out, she was happy to give the enterprising scout a sample of powdered barium sulfate for his latest project.

Back at the potting shed, David heated the barium sulfate in a frying pan until it liquefied. He mixed it with a small amount of the radium paint and a few pinches of radium-paint flakes, then strained the brew through a filter into a beaker. With uranium, David had bungled this very same procedure. Swift learner that he was, this time he judged the solubility of the two substances correctly. (It helped that he remembered to go to the hardware store beforehand for a proper filter.) The Geiger counter showed that the radium solution passed through to the beaker and the barium stayed behind. David took the beaker inside the house and quickly mounted the stairs to the bedroom. When he turned off all the lights in the room and lowered the shades, he saw that the liquid at the bottom of the beaker was emitting a faint glow.

Using a simple technique he'd first read about in the *Golden Book*, David dehydrated the radium solution into crystalline salts. He intended to pack the salts into the cavity of a new neutron gun he hoped to build from a larger block of lead than the one he had used in his original americium model. Like the talented if reckless scientist he was becoming, though, David prepared a final test of the radium's strength.

He had read that Pierre Curie had once impressed his friends by carrying a clear vial of radium mixed with zinc sulfide, which causes radium to become even more luminous. After a few hours' research at a branch of the Macomb County Library, David emerged with a plan. The only major item it required was a color TV set, and he found a battered old model at a junkyard. He paid a few dollars for it, threw it in the back of the Pontiac, and drove to Golf Manor. There, he smashed open the screen with a hammer and pulled out the picture tube. Inside it was a light greenish powder—zinc sulfide, which illuminates the screen.

David feared that his radium salts would lose their alpha potency if he mixed them directly with zinc sulfide. To prevent that, he mixed a small quantity of radium and aluminum in the chamber of his neutron gun and channeled a stream of alpha particles at a pile of the zinc sulfide he had placed on a piece of ordinary writing paper. When he placed the paper under a microscope and looked through the eyepiece, the pile was emitting small, luminous sparks.

And now another piece of good fortune befell David. His old friend who worked at the Macomb Community College chemistry lab had, as a favor, ordered a small stock of beryllium for the lab and turned over a few strips to David for use in his neutron gun. With a block of lead, David finished building a new radium-and-beryllium-based neutron gun in place of his earlier americium-and-aluminum model.

He was now ready to target neutrons on his radioactive thorium—and, with low expectations, on the dregs of his uranium "milk shake" as well—in the hopes of producing at least some fissionable atoms. He placed the gun on the shed floor, put the thorium strips in a foil-lined bowl directly in front of it, surrounded the whole thing with the sheet of malleable lead— folded at the top to create a dome—and let it sit for hours while he did homework or watched TV. He repeated the procedure for a few days, and when he measured the results with his Geiger counter . . . it didn't work. Based on his readings, the thorium had apparently grown only mildly more radioactive, while the uranium remained a total disappointment.

Once again, Professor Hahn sprang into action, writing his old friend Erb at the Department of Energy to discuss his (strictly hypothetical) problem. The department's isotope specialist had the answer. David's neutrons were too "fast," a problem originally discovered by Manhattan Project scientists. Hence, David would have to slow his neutrons by using a filter of water, deuterium, or, best of all, tritium, which would enable the neutrons to better penetrate their targets of uranium and thorium atoms.

Consulting his NRC list of commercially available radioactive sources, David saw that tritium is found in night-vision gun sights. He spent about seventy-five dollars of his dwindling cash reserves to buy a tritium sight from a local hunting store. At his father's house, David put on latex dish-washing gloves and with a little wooden coffee stirrer scraped out the tritium, contained in a waxy yellowish substance inside the sight, then collected the small sample in an old perfume vial. David feared that the tritium content would dissipate, but that was not the case—as he inadvertently discovered in a painful and dangerous fashion. That night, David took off his watch before bedtime and laid it on a table, directly atop the waxy substance on the coffee stirrer, which he'd absentmindedly forgotten to dispose of. He put the watch on before heading to school the next morning and by the end of the day noticed a painful burning sensation above his wrist. Taking off his watch, he found a small black radiation burn and realized what had happened.

David was only mildly concerned about the burn. Tritium, he reasoned, emits beta particles, which can penetrate only one or two centimeters of human flesh (unlike gamma rays, which have great penetrating power and can irradiate the entire body). He simply treated the injury with an antibiotic cream, which wouldn't have done much good other than alleviate the pain, and proceeded apace.

David had determined that his tritium was potent, but he had only a tiny amount. To economize his cash reserves, he began ordering sights from mail-order catalogues and sporting-goods stores, scraping out the tritium, and returning the sights to the seller, indignantly claiming that the product was defective. His ruses were not foolproof. One day, Ken received a call from the irate owner of a local gun-and-ammo shop. It seemed that David had asked to borrow a number of gun sights for a "demonstration" at Chippewa Valley. Following the demonstration—which of course never took place—David returned the sights, but the storeowner soon realized that they no longer worked. He didn't know exactly what David had done, but he forced Ken to pay for them.

Ken sat David down for a stern talking-to. Given David's history, he might have pressed him as to why the sights were defective and sought to discover what his son was up to. Instead, he simply demanded that David repay him and established a plan that called for four equal installments of one hundred dollars.

It was a crippling financial blow to David, but by now he'd accumulated enough tritium to proceed with his experiments. He took a smear of the waxy substance and dabbed it over a beryllium strip, which he reloaded into his neutron gun with the radium salts. As before, he placed the various components on the shed floor, imperfectly encased in the lead dome, and left them there until he returned, sometimes in a few hours and sometimes the next day. He carefully monitored the results with his Geiger counter over several weeks, recording his observations in his logbooks. To his joy and amazement, the thorium seemed to be absorbing neutrons: With each reading, the powder and strips were growing more radioactive.

David had succeeded in re-creating many of the same experiments that Marie Curie and other nuclear pioneers had performed during the first half of the twentieth century. Now, with the end of his junior year of high school approaching, he would do in his backyard what corps of scientists had worked decades to achieve: assemble a breeder reactor. The night that he reached this resolution, David was sitting alone at the patio table in the backyard in Clinton Township. "God, please let me finish this," he said as he stared up at the star-filled sky. "I will give anything."

His timing was perfect. With the summer at hand, David could devote himself entirely to the challenge, with the exception of the time spent at the busboy jobs he worked to finance his obsession. After clocking out, he could be found reliably in the potting shed, where he was methodically putting the final touches on his model breeder. His initial step was to take the highly radioactive radium and americium out of their respective lead casings. After a round of filing and pulverizing, he mixed those isotopes with beryllium strips and aluminum shavings, all of which he wrapped in aluminum foil, thinking the latter would intensify the production of neutrons. What were once the neutron sources for his gun became a makeshift core for his breeder reactor.

To contain his radioactive thorium strips and powder, he made dozens of tiny foil-wrapped, bouillon-style cubes, a tedious task that took the better part of a week. Thinking back to the checkerboard model he'd seen in his father's textbook, David built the breeder's outer layer with half-inch cubes that contained thorium-dioxide ash alternated with cubes filled with carbon. He selected the latter after reading in *Modern Chemistry* that neutrons multiply upon hitting carbon. Closer to the center came still more foil-wrapped cubes, these about one-eighth inch on a side and filled with a mixture of thorium ash and David's meager supplies of uranium powder. He assembled the cubes, the contraption held together tenuously with duct tape and foil around the neutron-generating aluminum-ball core. His completed breeder was the size of a shoe box and weighed about two pounds.

David didn't know if his breeder would work, but he felt that he'd achieved something pretty spectacular. "I never really believed I could get that far," he remembered. "I felt great." If David had worked with partners or had a professor overseeing his work, this would have been a moment for backslapping and congratulations, perhaps a night on the town for a celebration. But, as always, he was alone in the shed. There was no one to commemorate with, nor even anyone to call. Instead, after a few moments of admiring his creation, he pulled out his Geiger counter and recorded his readings in a logbook.

David continued to monitor his breeder reactor at the Golf Manor laboratory with his Geiger counter and record his observations. "It was radioactive as heck," he said later. "The level of radiation after a few weeks was far greater than it was at the time of assembly. I know I transformed some radioactive materials. Even though there was no critical pile, I know that some of the reactions that go on in a breeder reactor went on to a minute extent."

In fact, even if incapable of sustaining a fission reaction, David's model power plant was emitting plenty of radiation. His Geiger counter even detected radioactive emissions from his breeder through one of the inch-and-a-half-thick concrete slabs that he'd found in the potting shed when he took it over and had now retrieved from the bottom of the pool where he had dumped it.

David's first thought was that if he rigged an on/off switch for his reactor—a minor problem he'd thus far overlooked, never really thinking he would need one—he might be able to control and regulate the reactions inside the breeder. There was no time for Professor Hahn to write away for help; by the time he got a reply in the mail, his reactor might be the atomic equivalent of a runaway train. The next morning, David cornered Jim Miller at the school-bus stop and called him away for a private conversation. The words poured out as he brought him up to speed on his activities, but as he neared the end of his summary he hesitated. There was one little problem, he confided to Jim: The breeder was generating too much radiation.

Jim knew just what to do. Real reactors, he reminded David, use control rods to regulate nuclear reactions. He suggested that David fashion a makeshift scram system with cobalt, which absorbs neutrons but is not itself fissionable. "Reactors get hot, it's just a fact," Jim later said with a world-weary sigh in discussing his role in the breeder affair. At the time I interviewed him, David's former nuclear consultant was working as a cook at a Burger King in Clinton Township. As he munched on a bag of French fries, he explained that he was a self-taught nuclear specialist much like David, picking up information through independent research and by watching "informative TV shows."

Following his talk with Jim, David went to a local hardware store and bought a set of eight cobalt drill bits. He wrapped wire around the bits, pulled a cube out from the top of his model reactor, and inserted the bits between the core and the thorium-and-uranium cubes. But the cobalt didn't produce the desired result. His Geiger counter was detecting radiation emissions from a growing distance, and David feared events were spinning out of control.

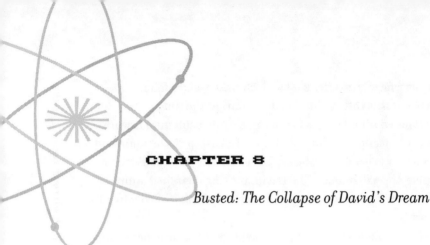

CHAPTER 8

Busted: The Collapse of David's Dream

WASTE DISPOSAL. If you can dump your waste directly into the kitchen drain (NOT into the sink), you are all right. If not, collect it in a plastic pail to be thrown out when you're finished.

—THE GOLDEN BOOK OF CHEMISTRY
EXPERIMENTS, 1960

For years, David had blithely worked with deadly radioactive materials without any appreciation of the risks he was taking. Now, for the first time, he was frightened by what he was doing in the potting shed. Still, he clung to the hope that he was overreacting. David took his Geiger counter to Chippewa Valley and convinced his electronics teacher to recalibrate it, telling him that he thought it might be defective.

After retrieving it the next day, he took the Geiger counter to his father's house and turned it on in the backyard. He registered nothing more than normal background levels of radiation—about fifty counts per minute (CPM). (Eighty-two percent of background radiation comes from natural sources, mostly radon, an odorless, colorless gas given off by natural radium in the earth's crust. The other 18 percent comes from man-made sources, such as nuclear-power plants, medical labs, and consumer products—like gas lanterns with thorium-dioxide mantles.)

Hopping into his Pontiac, David drove out to Golf Manor. When he pulled into Michael and Patty's driveway, he turned the Geiger counter on and started walking toward the backyard. Even before he got there, the counter started clicking, registering radiation far above background levels. The problem, it was clear, was not with the Geiger counter; it was with the model breeder reactor.

David could no longer ignore the fact that, sustained reaction or not, his breeder could be putting others in danger. Then there was the issue of what he might be doing to himself, a question that took on particular urgency when during precisely this time he stumbled upon the story of the radium-dial women, who had died terrible deaths from cancer brought on by ingesting Undark paint. This was normally the sort of "negative" reading David would have swiftly put aside, but he found the tale compelling—and utterly terrifying. He could vividly imagine the young girls' excitement as they covered their teeth and fingernails with radium paint, only to be consumed later by its radioactive contents. As he recalled, "I knew in theory that radium could be dangerous, but I'd never read about what had happened to people who handled it. I had no clue of any of that. I was pretty much scared to death."

So a few days later, with his Geiger counter picking up radiation five doors down the block, David saw no choice but to oversee an emergency shutdown of his baby reactor. Given the makeshift construction, the process didn't take long. Wearing his gas mask and gloves, David disassembled the breeder and separated its component elements into various containers. He took the thorium ash from the reactor's fuel cubes and returned it to its shoe-box container, which he stashed in his bedroom closet. He put the radium, americium, and beryllium from the core in individual glass jars and stored them and a few milk jugs of thorium ash in the basement. He packed most of the rest of his equipment, along with some of the reactor cubes that he didn't have time to take apart, into the trunk of the Pontiac and headed for his father's home in Clinton Township. With the object of his long-standing obsession literally in pieces, David was supremely dejected, but he still harbored hopes of picking up his nuclear threads at a later date. For now, though, he'd have to spend a few days dealing with the current mess.

As it turned out, David didn't have a few days. At 2:40 A.M. on August 31, 1994, less than twenty-four hours after he'd dismantled his laboratory, the Clinton Township police responded to a call concerning a young man who had been spotted in a residential neighborhood, allegedly stealing tires from a car. When two officers arrived on the scene, they found David sitting in his parked Pontiac.

It never became clear what exactly David was doing in his car at that late hour. He told the police he was waiting to meet a friend. Unconvinced, the cops decided to search his car. They opened the trunk, expecting to find the stolen tires; instead, they were confronted with an altogether more disturbing spectacle. The trunk didn't contain any stolen car parts, but it did hold a red toolbox sealed shut with duct tape and a variety of weird objects: fifty foil-wrapped cubes containing a mysterious gray powder, small disks and cylindrical metal objects, lantern mantles, mercury switches, a clock face, ores, fireworks, vacuum tubes, and assorted chemicals and acids.

David was calm but uncooperative. He warned the bewildered and alarmed officers that the gray powder in the cubes was radioactive but otherwise refused to provide any explanation for the trunk's strange set of objects. Even in these relatively halcyon days before Osama bin Laden became a household name, the cops wondered whether they had a teen terrorist on their hands. They filed a report at the time stating that the toolbox was "being treated as a potential improvised explosive device" and they feared it contained radioactive materials. In other words, the toolbox might be an A-bomb.

For reasons that are hard to fathom, the officers ordered a car containing what they feared was a nuclear bomb to be towed to police headquarters. "It probably shouldn't have been done, but we thought that the car had been used in the commission of a crime," Police Chief Al Ernst later said sheepishly. "When I came in at 6:30 in the morning it was already there." With a potential atom bomb parked in their lot, the police were forced to cordon off part of the area surrounding the headquarters.

David was thrown into an empty jail cell while the police

tried to determine what, if anything, to charge him with. How, they wondered, had David obtained the radioactive materials in the foil cubes? The police feared that he might have stolen them from a local company or that he had a collaborator at an atomic facility, like Detroit Edison's Enrico Fermi II, a conventional nuclear-power plant in nearby Monroe. Of course, the most pressing issue was establishing whether they were dealing with an A-bomb.

Meanwhile, David's long-suffering father and stepmother were once again surprised and horrified to discover that their son had been carrying out scientific activities behind their backs and that events had, once again, run seriously amok. Early that morning, before the sun had fully brightened the suburban sky, Ken received a testy phone call from a detective who told him to get down to police headquarters as quickly as possible. When he arrived, the detective unceremoniously escorted him to an interview room and told him that the police believed it possible that David was building an atomic bomb. The detective demanded to know what the father knew about the son's activities and seemed dubious when Ken told him that he was as mystified as the police were.

Ken's obliviousness about his son's atomic freelancing was probably heightened by his inner makeup and his hometown's psychology. In Ken's world, research and innovation were everyday pursuits, as were the engineering and construction projects they led to. Indeed, Detroit—the Motor City—lionizes the bootstrap inventor who discovers new methods of industrial production and manufactured products. David would have attracted hostile attention had he joined up with an

environmental group with a battle cry of "Parks, Not Roads";
pursuing a new form of energy was par for the course.

The police brought David from his cell to the interview
room. Ken was fuming. David had duped him, lied to him, and
generally made a fool of him. Worst of all, he'd taken advantage
of Ken's sentimental love of scouting. Whenever his father asked
him to explain his strange doings, David had told him that they
were for upcoming demonstrations for Troop 371, and Ken was
constitutionally unable to be critical of the scouting enterprise.

As his father stewed, David remained largely uncooperative
and taciturn, which made the investigation difficult. The police
realized that they were going to need outside help, as nuclear
physics was beyond the station's normal call of duty. Chief Ernst
phoned the Michigan State Police Bomb Squad to examine the
toolbox and called in the state Department of Public Health
(DPH), located ninety miles away in Lansing, to supply scientific
and technical assistance. Dave Minnaar, who heads the
radiological division at the DPH, quickly dispatched a team from
the agency's division in nearby Pontiac.

The DPH crew arrived along with the bomb squad, which
went to work first. Before long, they had good news to report.
The toolbox was not an explosive device. After making that
determination, a bomb-squad crew member opened it up and
found the same hodgepodge of foil-wrapped cubes, chemicals,
and doodads that had been floating around in the Pontiac's
trunk.

Now the DPH took over. An employee wearing protective
gear ran a radiation meter over material from the trunk and
toolbox. His news was less encouraging. There was no cause for

panic, but some of the materials were very radioactive. The DPH gathered the materials into secure containers and began to run a series of tests. Within a few hours, the results were in: David's trunk contained materials with high concentrations of americium and, far worse from the point of view of public health, levels of thorium "not found in nature, at least not in Michigan."

This discovery automatically triggered the Federal Radiological Emergency Response Plan. The FRERP was originally developed after the Three Mile Island accident by a task force of sixteen federal agencies. It is designed to enable the government to respond to a peacetime radiological emergency and execute an evacuation plan if necessary. In case of a code-red FRERP crisis, the government uses the Emergency Broadcast System to warn people to go inside their homes or public buildings, close the doors and windows, turn off heating and air-conditioning systems (which could spread contamination), and await further instructions. Law-enforcement agencies establish control points to speed traffic out of the affected area and block access into it. They may also set up a rumor-control center to issue information to the press and contain panic.

The unfolding situation at Clinton Township police headquarters was no Three Mile Island, but it still called for consultations with federal agencies in Washington. Soon, state officials were embroiled in tense phone conversations with the Department of Energy, the EPA, the FBI, and the NRC.

The first order of business, the authorities agreed, was to confirm that the Pontiac contained David's entire stock of radioactive wares. David assured them it did, but Minnaar ordered a team to Ken and Kathy's home to make sure. When readings taken inside and outside failed to uncover any traces of radiation, everyone breathed a sigh of relief.

Of course, Ken and Kathy's home hadn't been the site of any of David's advanced experiments; those had been conducted thirty miles away, at his mother's house in Commerce Township. And there, sitting in the basement and bedroom closet as all these events were transpiring, were David's radium, significant quantities of foil-wrapped ash, and a good deal of other radioactive material.

But the police and DPH didn't know about the potting shed, nor did they know of Patty and Michael, who learned of David's arrest only months later. (Realizing it would alarm his fragile mother, David never told her about his run-in with the police. It never occurred to the pathologically oblivious Ken—who was incapable of seeing that Patty might have useful information about, or insights into, their son's weird behavior—that he should do so.) The authorities thought Kathy was David's birth mother, not his stepmother, and that Clinton Township was his sole residence. Therefore, they believed everything was under control. The cops were still furious with David and considered charging him with illegal possession of explosives, but there were no real grounds to hold him. The Pontiac's trunk contained some dangerous items but not a bomb. And so, at the end of the day, the police released David to Ken's custody.

Ken and Kathy had planned to leave a few days later for Torch Lake in northern Michigan, where they kept a twenty-four-foot travel trailer on a lot near Traverse City. Not trusting David to be left at home, they took him with them and pitched a tent outside the trailer for him to sleep in.

The story of David's radioactive experiments might well have ended here, if not for the fact that Minnaar wanted to run a few more questions by the would-be breeder builder, largely to satisfy his own curiosity. Minnaar was amazed that a teenager had been able to secure an element as tightly controlled as thorium, and he suspected that the foil-wrapped cubes and other materials found in the trunk were part of a larger plan that David was keeping from the authorities.

Hence, Minnaar scheduled a quick follow-up interview with David, only to postpone it when the DPH had to deal with an alert on the other side of the state, where high levels of radium had been found at a trash dump. It wasn't until Thanksgiving Day, about three months after David's arrest, that Minnaar found time to call him at his father's house. During their brief conversation, David said he'd been trying to make thorium in a form he could use to produce energy and that he had hoped "his successes would help him earn his Eagle Scout status." He insisted that he'd been very careful and that the authorities had no reason to worry about any environmental contamination stemming from his activities.

David might have believed he'd been careful, but Minnaar didn't for a second. It was obvious from their conversations that David knew a lot about the basics of radioactivity, but it was equally clear that he didn't know much about protecting himself from it or even how much represented a serious risk. Unsettled, Minnaar called David again a few days later. Ken or Kathy picked up the phone—Minnaar can't remember which—and casually said that David wasn't in; he was at his mom's house in Commerce Township.

His mom's house?

This was the first Minnaar had heard about Patty. He immediately called David at Golf Manor and demanded to know if he had conducted any radioactive experiments in Commerce Township. It was only now that David, figuring that Minnaar was going to find out anyway, confessed about his backyard laboratory. He also offered a fuller accounting of his experiments, though, fearful as he was of legal repercussions, it was still a heavily censored version. But he said more than enough—especially about the method he used to purify thorium and how he used a blowtorch on the lantern mantles—to astonish and appall Minnaar. "[I] suspected that he was pursuing the development of a new energy source involving nuclear fission," the DPH man said later. "He talked about this as his goal, and he knew enough about radiation and chemistry that [I] couldn't rule it out."

In early December, state radiological experts arrived without warning in Commerce Township to interview Patty and Michael and survey the potting shed. Patty and Michael were watching TV in the living room when the team knocked on the door. Now they finally heard of David's arrest and the strange things that the police had found in the trunk of his car. "They asked if we had known what David was doing, and we said no, we sure didn't," Michael remembered. "They told us he was trying to build a nuclear reactor! We couldn't believe it!"

After finishing the interview, the radiological team went to the backyard, where they found the shed in a frightful state, covered with spiderwebs and strewn with junk. Having determined after his arrest that it would be best to lie low for a while, David hadn't been there for months. An old, broken air conditioner was stuffed into a corner, and there were burned pans filled with ashes littering the floor, along with jars of acid, toppled-over milk crates, and kicked-over coffee cans. Much of the shed was contaminated with what subsequent official reports would call "excessive levels" of radioactive material, especially americium-241 and thorium-232.

Minnaar was relieved that only a small level of contamination was found outside the shed. There were no signs that the neighborhood had been dangerously irradiated. On the other hand, the extent of the contamination and the purity of the thorium remains found in the shed surprised him. That anyone, let alone a kid, had gotten his hands on such potent radioactive substances made the whole episode unique in the DPH's experience. In a typical incident handled by the department—indeed, until this point the only type of incident—radiation escaped from a *regulated* facility during the course of a *regulated* activity. In Golf Manor, the DPH had the ultimate unregulated situation: a teenager producing and working with homemade sources of radiation. As Minnaar later said, "These are conditions that regulatory agencies never envision. It's simply presumed that the average person wouldn't have the technology or materials required to experiment in these areas."

The DPH crew tested a vegetable can found in David's shed—it had been used to handle thorium and radium, though that wasn't known at the time—and got a reading of 50,000 CPM, about one thousand times higher than normal levels of background radiation. A copper bowl registered at 6,000 CPM, a paper scrap at 3,000 CPM, and a solid black clump on the shed floor at 1,500 CPM. At the lower end of the spectrum, a "black spot on a triangular concrete chunk" came in at 300 CPM, a pickle jar at between 160 and 180 CPM, and a blackish-grayish dust found in a bottle of Gilbey's vodka registered at 90 CPM.

Under federal regulations, contamination levels exceeding 1,000 CPM are considered unacceptable at a residence. Levels can be higher at a workplace but generally still can't exceed 20,000 CPM. That type of reading might be found at a hospital nuclear-medicine lab, where radiopharmaceuticals are prepared for patient use. But a regulated workplace where radioactive materials are handled requires a full range of safety measures, none of which was found in David's potting shed. For example, only trained personnel in protective gear are authorized to work in a nuclear lab. Detection meters are posted to monitor the air, and strict rules are in place so that the slightest radioactive contamination can be controlled and swiftly eliminated in the event of an accident. Every piece of equipment in a regulated workplace is designed to prevent the spread of radiation. Lab tables, for example, are made with special nonporous surfaces that are covered with absorbent materials to soak up spills.

What's truly scary is that history will never record the true radioactive levels of David's laboratory. After Minnaar spoke with him at his mother's home, David told Michael and Patty that the EPA might need to talk to them about his experiments. He neglected to mention his arrest, but he said enough about the situation to scare the couple. Petrified that the government would take their home away as a result of David's activities, Michael and Patty searched the house and gathered up the materials that David had hidden the day he disassembled his reactor; when Minnaar's team arrived, they had already been disposed of. "I'm a truck driver," Michael later said. "I didn't make a great deal of money, but I got a nice house. We thought they were going to take everything away. I had to drink Nyquil to get to sleep."

Michael and Patty discarded in the household trash most of what they found, including David's radium, pellets of thorium, and several quarts of miscellaneous radioactive powders. Those materials, especially the radium and thorium, were surely far more radioactive than anything found by state officials—and ended up dumped at a local landfill. To which David remarked, "The funny thing is, [the state surveyors] only got the garbage, and the garbage got all the good stuff." State officials, who didn't learn of this at the time, would doubtless have been less amused.

Having determined that no radioactive materials had leaked outside the shed, state authorities sealed it with a padlock and ran yellow warning tape around it to keep people out, especially David (whom they urged to consult a doctor about potential health damages he may have suffered due to his experiments). Then they petitioned the federal government for help. The NRC would have seemed to be the logical choice since it licenses nuclear plants and research facilities and deals with any accidents at those sites. But David was not an NRC-licensed operation, so it was determined that the EPA, which responds to emergencies involving lost or abandoned atomic materials, should render assistance. An assistant of Minnaar's fired off a memo to Richard Karl, chief of the EPA's emergency-response branch in Chicago, laying out the scope of the radioactive problem and pleading for assistance. "The extent of the radioactive material contamination within a private citizen's property begs for a controlled remediation that is beyond our authority or resources to oversee," he wrote.

The EPA didn't need much convincing. Jim Mitchell, an official at the Chicago branch, said that his office was stunned upon learning of David's thorium stockpile. "We had a very big problem with that," he said later. "It was some of the highest concentrated material that we had encountered in private hands."

On January 25, 1995, officials from the Chicago EPA arrived in Golf Manor to conduct their own survey of the shed. An official assessment conducted at the time reported no noticeable damage to flora or fauna in the backyard. But forty thousand nearby residents had been put at risk during David's four years of experimentation by the potential release of radioactive dust and radiation from his hammering, soldering, and explosions.

The findings convinced Karl that drastic measures were required—so drastic that he wanted David's potting shed classified under the federal Superfund law, which gives the government wide authority to respond to potential and actual ecological disasters. Superfund sites run the gamut from a water table tainted by motor fuels all the way up to the work at the Rocky Flats plutonium production center. Superfund's biggest job ever was the cleanup of Prince William Sound in Alaska after the *Exxon Valdez* spilled eleven million gallons of oil there in 1989.

Karl's office explained its thinking in an "action memo" to Jodi Traub, associate director of the Superfund division in Washington. In it, the emergency-response branch warned that conditions at the shed presented "an imminent and substantial endangerment to public health or welfare or the environment" and that there was "actual or potential exposure to nearby human populations, animals, or food chain." The memo stated further that adverse weather conditions such as heavy wind or rain could cause the "contaminants to migrate or be released" and that the state "does not have the necessary resources to respond to this time-critical situation."

The memo spelled out in grim detail the consequences of inaction. If thorium dust in the shed was blown aloft, Golf Manor residents could face an increased risk of developing cancer of the lung or pancreas, not to mention chromosomal damage. Exposure to radium, Karl pointed out, could weaken the immune system and cause anemia, cataracts, fractured teeth, cancer, and death. Traub needed only a brief period of reflection before she ordered a cleanup under the Superfund law.

In June 1995, the EPA deployed the moon-suited cleanup workers to Golf Manor. The crew had been told that a teenager had been conducting radiation experiments, but they were utterly mystified upon forcing open the lock on the potting shed. The lab's equipment "looked like stuff that the kid bought at a grocery store or salvaged from a junkyard," said Barclay Albright, a member of the crew. "He had tin cans, silverware, bowls, and other stuff in there, plus car parts and other scrap that was thrown all over the place."

Even more bizarre, David had painted bold red warning

signs on the shed's inner walls, like "CAUSHON" and "RADIOACTVE." "We were amazed with what he'd done, but we couldn't make sense of it," Albright continued. "Here's a kid who's using high school textbooks to re-create Madame Curie's experiments and doing it well, but he can't pass a spelling test."

The crew was dying to meet David, but he was determined to keep a low profile and spent the three-day cleanup period at his grandma Lucille Spaulding's home. The adults in David's life were also missing in action. Michael and Patty, still worried about having their home seized, hunkered down inside as the crew worked. Ken and Kathy didn't know about the scheduled cleanup, as David, hoping it might glide by unnoticed, hadn't said a word about it, and they had never followed up with Minnaar. In a phone call with her former husband, Patty casually mentioned that the EPA planned to come by to inspect the house but said nothing about a full-scale assault on the potting shed.

Hence, when Albright and his crewmates descended on Golf Manor, Ken and Kathy were enjoying a ten-day vacation at a state park near Clare, Michigan. They would normally have taken David and Kristina with them, but Kathy, her nerves frayed from the past year's traumas, suggested that the two of them go alone.

The only spectators at the cleanup, at least its first day, were Dottie Pease and other nervous neighbors. The crowd expanded over three days as EPA officials infuriated Golf Manorites by refusing to provide solid information about what they were doing on Pinto Drive. After Dottie called her husband at work and hysterically warned him to "do something," Dave Pease sped home from the office. When he arrived, cleanup supervisors remained tight-lipped, telling Dave that the crew would be done before long and that he and his wife should go home, enjoy a good dinner, and stop fretting.

But it's hard to enjoy your dinner when men in moon suits are noisily sawing down a potting shed a foot away from your back fence. Angry and frightened, Dave alerted the local press. By the next morning, TV stations and newspapers had deployed camera crews and notebook-wielding reporters to Golf Manor. That finally forced the EPA to start talking, though officials still offered only a narrow and misleading account, stating that a teenager had been tinkering around with radioactive materials found in household products. The agency refused to give David's name to reporters, and though a few learned it from neighbors, neither he nor any family members agreed to be interviewed.

For these reasons, the subsequent news reports barely scratched the surface of the story. One local newspaper, *The Oakland Press*, explained to its readers that the men in funny suits were called in response to "a burgeoning teenage scientist [and] his makeshift radioactive laboratory." The story said that the boy had "gathered the materials from radioactive rocks, gas lamps and about 100 smoke detectors" and that he was "chemically concentrating" the materials for unknown purposes. Another local paper reported that "an enterprising teenager" had followed in the footsteps of Pierre and Marie Curie—a description that must have pleased David—by "using powerful acids to extract tiny particles of radioactive materials from household items." The farthest David's fame spread was to the *New York Post*'s Page Six column, which is normally reserved for items about the latest doings of local socialites and gossipy tidbits about Hollywood celebrities and fashion models. "Most people keep rakes, shovels and lawnmowers in their garden sheds," began the item, which ran to just a few lines. "But not 'Dabbling' David Hahn," who stored radioactive materials there until the EPA "discovered the stash and carted it away."

John Sims, David's stepgrandfather, was at his summer house in northern Michigan when the cleanup took place. He was reading his newspaper over breakfast when he saw an item about an unnamed teenager near Detroit who had been conducting strange experiments in his backyard. Having moved to a new home farther from Clinton Township, John hadn't spent as much time with his stepgrandson in the past few years. Even so, he recognized the fingerprints. "I turned to my wife and said, 'This has got to be David,' " he recalled with a trace of admiration, though perhaps more distress. "I knew it in my gut the second I saw the story."

After their third day at Golf Manor, the cleanup crew had fully dismantled the potting shed. The wood planks that once comprised it and everything found inside had been deposited in thirty-nine sealed barrels marked with the radioactive symbol. In the late afternoon of June 28, a semitrailer pulled up to the curb in front of Michael and Patty's house. Cleanup-crew members began hauling the barrels from the backyard and loading them onto the truck. When the final one was aboard, the truck and its two occupants gave a wave to the crew and the scattered band of Golf Manorites on hand and headed out of the neighborhood.

The nation's most highly radioactive refuse—a depleted nuclear-power-plant core, to take one example—is buried in sealed repositories that are designed to last for thousands of years. The bulk of such refuse is composed of mineral ores, contaminated soils, construction material, and other items that are less deadly but still too "hot" to toss in a common trash dump. David's potting shed fell into this last category, and so the EPA sent the semitrailer to Envirocare, the sanitized name for a dump facility—a square-mile trench lined with thick plastic and capped with an impermeable claylike soil—located some 1,700 miles away in the middle of the Great Salt Lake Desert in Utah. There, the remains of David's experiments, minus what ended up at the landfill in Michigan, were entombed, along with tons of radioactive debris from the government's atomic-bomb factories, plutonium-production facilities, nuclear-power plants, uranium mines, and contaminated industrial sites.

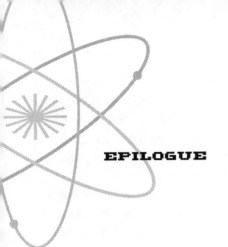

EPILOGUE

Postnuclear Syndrome:
David's Radioactive Exile

The newspapers have published numerous diagrams, not
very helpful to the average man, of protons and neutrons
doing their stuff. . . . But curiously little has been said, at
any rate in print, about the question that is of most urgent
interest to all of us, namely: "How difficult are these things
to manufacture?"

—GEORGE ORWELL, "YOU AND
THE ATOM BOMB," 1945

David went into a serious depression after the federal
authorities shut down his laboratory. He'd depleted all of
his savings—more than eight thousand dollars—on his
experiments, and he had little to show for his time and effort.

Years of painstaking work had been thrown in the garbage or buried beneath the sands of Utah. To make matters worse, during their frenzied cleaning spree prior to the EPA's arrival in Golf Manor, Patty and Michael had tossed out David's logbooks along with his radioactive raw materials. Hence, David's only record of his experiments was lost forever.

Despite Michael and Patty's fears, the EPA neither took their home nor sent a bill for the cleanup. It came to sixty thousand dollars, most of which went to cover transport and disposal in Utah. The EPA could, in fact, have billed them. Superfund's budget comes from taxes on oil and certain chemicals and from fees imposed on responsible parties.

A few days after the cleanup crew completed its job, Ken and Kathy returned to Clinton Township, tanned and rejuvenated from their sojourn at the state park. Their high spirits didn't last long. Almost as soon as they'd pulled into the driveway, several neighbors who'd seen news accounts of the cleanup came scampering from their homes to tell them about their son's newfound fame. They also warned them that TV crews had spent several days parked on their front lawn, hoping they might be able to snare an interview with—or at least catch a glimpse of—the parents of the Radioactive Boy Scout.

When David arrived home a few hours later, a predictably upset Ken demanded an explanation. For three hours, as his father listened in amazement, David gave a rambling account of his activities and how they had led to the EPA's intervention. "Dad, I'm on to something really big, a new way to produce energy," he told him. "You're not going to be able to understand all of it."

Even Ken couldn't ignore a transgression the size of a breeder reactor. He grounded David for two weeks and took away his car keys. He was to return home after school—David had started his senior year at Chippewa Valley just a week after the police learned of his breeder—and stay there, with no experimenting! Meanwhile, Ken, Kathy, and David began a short term of family counseling.

When Ken returned to work—he was then employed at a company that made robotic welding lines for cars—his boss, John Hergott, came over to his desk first thing in the morning. "I didn't know your son was a mad scientist," he said. "Why are you keeping him a secret?"

"What are you talking about?" Ken replied. "You don't even know my son."

Hergott pulled out a clipping from a local news account of the cleanup and laughed, saying, "Now I do—and so does half of the city!"

Hergott, the head of a local astronomy society and an amateur scientist, wanted to meet David, so Ken brought him in. The pair talked privately in his office for two hours and when they finally emerged, Hergott told Ken, "I've studied chemistry for years, and I couldn't tell him anything he didn't already know." Years later, with the passage of time, Ken spoke of Hergott's comments with enormous pride. At the time, though, his words left Ken with a feeling of dread and uncertainty.

Damage from radiation accumulates over time because once it's in the body it stays there. The question of how much radiation poses a risk is the subject of debate, but over the years the government has repeatedly lowered what it deems to be the acceptable level. In the aftermath of the cleanup, the EPA advised David to have a full body count. During that procedure—first performed on a group of people in the case of the radium-dial painters—a sensitive radiation detector is run across the body. David agreed, and the EPA arranged for him to undergo an examination at the nearby Fermi plant, which remained open as a research plant after its failed life as a breeder reactor. At the last minute, fearful of what he might learn, David backed out.

Other than the visit to Hergott, Ken granted only one other exemption from David's housebound status. David had still not completed the patio for the public library, the final requirement for Eagle Scout, and his eighteenth birthday—the deadline to complete all Eagle Scout requirements—was now less than two months away.

With his lab shut down, David had plenty of time on his hands. He spent the next few weekends toiling away, helped occasionally by Heather and his dad. In late September, with just a month to spare, David poured the final section. That night, Ken took him out for dinner to celebrate.

With the patio built, David sent a formal request for Eagle status to scouting's national board of directors. "My life purpose is to lead a clean and honest life that is both rewarding and peaceful, but full of new and exciting adventures," he wrote. "My primary ambition is to become a chemical engineer working with metals and new kinds of alloys." Soon afterward, Scoutmaster Auito notified him that he had been accepted into the Eagle's Nest.

David continued to take a breather from his radioactive

experiments even after his grounding ended. Perhaps he was relieved that the police had eventually forced matters to a head and brought a definitive end to his research, a step he might not have been able to take voluntarily. Perhaps, too, he recognized that he was in danger of losing Heather. For her, the arrest and its aftermath were more signs that David was not husband material. Their relationship dragged on, but its ardor was waning.

The newspaper accounts that appeared during the cleanup operation did bring David a small amount of flattering attention and recognition. A number of TV producers and national reporters were amazed by what he'd done and called him. Three employees of the cleanup crew later gave an awed presentation to a meeting of the Health Physics Society, a professional organization whose members specialize in occupational- and environmental-radiation safety.

At Chippewa Valley, teachers and students were stunned to learn that David's vague allusions over the years to nuclear research had been more than mere bluster. David made sure everyone knew about his story. He made hundreds of copies of the newspaper stories and passed them out as he walked the halls of Chippewa Valley. "I really liked giving the stories to kids who were always saying I was just making everything up," he said. "I'd walk up to them, hand them a story, and just walk away."

Still, his stature at the high school was little enhanced. Whereas a few friends had once affectionately dubbed him Glow Boy, he was now known more widely as Radioactive Boy. When for Valentine's Day Heather sent David balloons at the high school, they were seized by the principal, who apparently feared they had been inflated with chemical gases David needed to continue his experiments.

One day, the principal interrupted the class of Sue Young, whom David considered a mentor, and introduced her to two agents from the FBI. The agents wanted to know if David had gotten any of the chemicals or materials he used in his experiments out of her classroom. Young and the principal assured them that he had no coconspirators at Chippewa Valley. Neither Young nor the principal seems to have given any thought to offering counseling to David or otherwise trying to understand what made him tick. The entire breeder affair was kept under wraps while the school tried to protect its reputation.

Indeed, most adults looked upon David's experiments as an embarrassment and something best not talked about. No one concluded that he had talents that should be harnessed or that with a little more encouragement and guidance he might end up at MIT.

In a final indignity, some local scout leaders attempted to deprive David of his Eagle Scout status, arguing that his extracurricular activities had endangered the community. "It became very controversial," one scouting elder said of David's case. "You're supposed to follow the Scout Law, and there was a feeling that he'd done more than just cut a few corners. He lied, stole stuff, and put people in jeopardy." Ultimately, the move to strip David of his rank petered out, but the uproar heightened his isolation and woes.

And so, without direction or a mentor, David was left to drift.

As David's senior year of high school wound down, he descended into an especially bleak period, triggered by two devastating blows. The first was Heather's decision to break up with him. If David had been less distracted by his scientific endeavors he might have heeded some warning signs, but the decision blindsided him.

In later explaining her reasons for the breakup, Heather, then working toward a business degree at Grand Valley State University near Grand Rapids, said, "He had an on/off switch, and after about a year it was permanently on. I had no idea of what he was talking about most of the time. You'd have to look back to Einstein. These types of people are not made for relationships." David was crushed by Heather's rejection. He began following her around and would sometimes sit for hours in his Pontiac in front of the candy store where she worked—to the point that Heather got scared and cut off all contact, saying it was impossible to be friends.

A more profound and terrible shock came in February 1996 when Patty, roiled by mental illness, shot and killed herself in the kitchen of her home. Despite all of her problems, Patty had always showered her son with affection and support. For David, the only person who loved him unconditionally was gone, and the manner of her death made that fact all the more painful. He became more withdrawn than ever, finding it hard to carry on a conversation beyond a few mumbled words and at times worrying that he was losing touch with reality. "When my mother and Heather left, I felt as if I was being punished by God," David said. "All hope vanished from my life."

During the summer after graduation, David spent much of

his time in a fog, lying in bed or driving his Pontiac in circles around the block. In the fall, Ken and Kathy pushed him to enroll in Macomb Community College. He majored in metallurgy but skipped many of his classes and continued to spend most of his time at the house, absolutely despondent. Finally, his father gave him an ultimatum: Join the armed forces or move out on his own. Ken called the local navy recruiting office, which sent a representative by the house. Sullen and depressed, David gave in. In late 1996, he shipped out to boot camp, and three months later was ironically posted aboard a nuclear-powered aircraft carrier, the USS *Enterprise*.

David had a hard time adjusting to life aboard *Enterprise*, which is home to more than five thousand sailors and sometimes spends months at sea before returning to its base in Norfolk, Virginia. Word of his radioactive experiments leaked out to shipmates, who promptly nicknamed him the Nuclear Kid. David had hoped to further his atomic education in the navy, but ship officers, in his words, "[wouldn't] let me near the nukes." He was barred from entering an engineering training program and to his great dismay had his request for a tour of the ship's reactors turned down. Alas, as a lowly seaman, David's duties aboard *Enterprise* were largely of the deck-swabbing and potato-peeling variety. With no real home, no privacy, and no chance of doing research, David's spirit was crushed.

It may seem a shame that David's scientific talents were lying fallow, but it was perhaps best that he took a respite from the nuclear field, for government scientists still worried that his previous exposure to radioactivity may have seriously shortened his life span. Thorium emits relatively weak radiation, and therefore poses little danger outside of the body. But if inhaled or ingested—a real possibility in David's case because of the way he was grinding and burning it in a cramped space while protected by inadequate safety gear—it greatly increases the risk of cancer. Neutrons are very penetrative of human tissue and difficult to control in a real laboratory, let alone when using a homemade neutron gun in a potting shed. Most troubling of all is the radium to which David may have been exposed. One section of his old notes that escaped destruction contained his transcription from an encyclopedia that said that radium can "migrate to the bone marrow, where [its] radiation interferes with the production of red blood cells. Less than one-millionth of a gram can be fatal."

Some three years after the cleanup was completed, I drove from Detroit to Lansing, where Dave Minnaar works in a dreary building that houses several state environmental agencies and is, in addition, the repository of birth and death certificates. Because Patty Hahn had cleaned out the shed before Minnaar's men arrived on the scene, he did not know that David had built a neutron gun, nor did he know about the vial of radium paint David had found in the antique clock and purified with barium sulfate. Nor did he understand, until I told him, that the cubes of thorium and uranium powder found by police were the building blocks for a model breeder reactor. After being filled in on David's full range of activities, Minnaar was shaken. "This is getting scary," he said. "The steps you describe, if followed, lead one to another. There's nothing I'd object to with his physics."

Of course, that doesn't mean that David had come anywhere close to building a working breeder reactor. He apparently succeeded in penetrating with neutrons the atoms of his thorium-232, but he hadn't turned it into fissionable uranium-233. Even if he had, he still didn't have the quantity or the quality of ash to sustain a chain reaction. Nonetheless, Minnaar was amazed that David had proceeded as far as he did down the nuclear path. Indeed, when he spoke about David's model reactor, he referred to it as if it were the real thing. "It would have been pretty hard to fine-tune the thing, but the evidence shows this guy knew his stuff," Minnaar said. "It appears he was working on a form of nuclear-energy production."

And what of David Hahn? I first met him in 1998, three years after the potting shed had been dismantled, at a McDonald's on the sprawling navy base in Norfolk where at age twenty-two he was stationed. Having heard about his exploits, I was anticipating some physical manifestation of brilliance or obsession, the powerful presence of an Einstein or a Unabomber. But all I saw was a beefier version of the clean-cut kid from childhood photos, with a peach-fuzz mustache, gangly limbs, and a muscular build from navy drudgery.

David's manner was oddly dispassionate, though polite, until we began to discuss his nuclear adventures. Then, the on/off switch Heather talked about clicked on and stayed on for the next five hours. Lighting and grinding out cigarettes—a habit he picked up on *Enterprise*—David enthused about his old research and the glorious days he spent in the potting shed. He was mildly embarrassed but not contrite about the thieving, deception, and destruction, saying, "I wanted to make a scratch in life. I've still got time. I don't believe I took more than five years off of my life."

I had more opportunities to meet David during the next few years, mostly in Detroit during his leaves from the navy. He stayed with Ken and Kathy at their new home, a two-story ranch house located in a neighborhood called Kingswood Estates. He'd finally gotten over Heather, though that had taken an unusually long time, probably because he hadn't had a serious romance after they'd split up.

During one visit, Ken, to David's obvious embarrassment, pulled out of a basement closet several boxes containing his son's old scouting memorabilia, including a cardboard sheet to which his merit badges and medals had been glued. One of the boxes contained a three-ring binder, where David's scouting progress reports had been stored neatly.

As a teenager, David's one-dimensional pursuit of his arcane goals left him virtually oblivious to world affairs and current events. As a young adult, his lack of awareness of the world he lived in had reached absurd proportions. During one conversation, when David was listing countries he has visited with the navy and others he'd like to see in the future, he blurted out, "Whatever happened to those communist countries? You don't hear much about them anymore"—this about a decade after the fall of the Berlin Wall and the collapse of the Soviet Union.

One day, David picked me up at my hotel for a predictably peculiar tour of the Detroit suburbs. He was carrying a new Geiger counter, which fit in his palm and—with a price tag of five hundred dollars—was far more sophisticated than the one he had mounted in his Pontiac. Our first stop was Gloria's Resale Boutique/Antique, the spot where he had found his prize vial of radium paint. David strolled through the store with his Geiger counter, testing a number of items as he went along. He turned up one old glow-in-the-dark clock with a dial coated with radium, though apparently only a small amount. David had better luck at our next stop, another antiques store that he customarily frequents when in Detroit. There he found a small electric clock, probably from the 1950s, which prompted a veritable spasm of clicking. David bought the clock for five dollars, and we were on our way again.

David didn't want to talk about his future plans and when pressed became coy and evasive. The closest thing to an answer came when he allowed me to sift through a selected batch of his research files. Based on what I saw, it would appear that he maintains regular correspondence with dozens of companies across the country. In one letter, a biologist at GNC LiveWell ("the nation's largest manufacturer and retailer of nutritional supplements," the letterhead crowed) expressed a polite sense of exasperation with David. "Thank you for taking the time to contact us again," the letter opened. "We enjoy hearing from our valued customers." Unfortunately, GNC could not help David with his questions about the oxidation of L-tyrosine to L-dopa, nor did the firm carry a product he was seeking, trivalent cysteine. In another letter, a customer-service representative at Fleischmann's Yeast in Fenton, Missouri, informed David that the company's products "do not contain the photolase enzyme" and suggested that he consult medical journals to find out more about it. (As consolation, the employee enclosed in the letter a complimentary sample of yeast and a collection of popular recipes.) David requested from DuPont brochures with opaque titles such as "Dark Secretions/Excretions" and "Melanoma (Cutaneous, Eye [Unveal])" and submitted an idea to Pfizer for a "tanning pill that produces or is a predecessor of melanin, a melanophore expander and a sun light protectant."

Tanning was, of course, one of his old obsessions, and there were plenty of signs, mixed in among his jumbled assortment of letters and papers, that David hasn't let go of his nuclear fixation either. His files bulged with information on what sort of ores are available at public mining sites around the world, including the open-pit Ruggles Mine in Grafton, New Hampshire. According to a colorful brochure from the mine, Ruggles boasts ample supplies of 150 minerals, including a type of uranium that is "prized by collectors and museums the world over." Then there was a letter from Horst Windisch, the liaison officer at a mining federation in South Africa, replying to David's request for information about the quality and availability of pitchblende at the Rössing Mine in neighboring Namibia. Windisch wrote that while the pitchblende there emits only a small amount of radiation, it is still a controlled mineral and hence he would not be able to send along a sample.

David also casually mentioned to me that he had recently contacted the NRC by mail and asked for information about uranium oxide and polonium. Whoever received the request seems to have recognized David's name—by now, punching the words *David Hahn* into the NRC's central computer system should probably trigger a systemwide security alarm—and notified David's commanding officer. David was "just trying to clear up some things from the past," he said, though that explanation didn't mollify officers aboard *Enterprise*. If he wanted to send any further letters that in any way touched on the subject of nuclear power, he was told that he would need prior approval from his commander.

After completing his four-year naval tour of duty at the age of twenty-four (leaving the service as an interior communications specialist with a rank of petty officer, third class), David returned home to Clinton Township in June 2001. He remained ardently pro-atomic and still believed, though perhaps a bit less contemptuously, that critics of nuclear technology were simply ill informed and lacking in sufficient technical knowledge to understand how it worked. David initially planned to get his undergraduate degree at Macomb Community College and then hoped to transfer to the University of California at Berkeley to pursue a postgraduate degree in nuclear chemistry. "This time," he said, "I'll do it the right way."

I spoke to David periodically to check up on his progress. Once, Ken answered the phone and told me regretfully that David was gone for a few days. "He's up in Canada. Mining uranium." The last two words were almost an afterthought, though they prompted in me the same sinking feeling Minnaar must have felt years earlier upon learning of David's potting shed. Sensing my reaction, Ken, eyes wide-shut as always, added hastily, "This is just small-scale stuff, nothing like before."

It turned out that David and a navy friend were hundreds of miles north, near Bancroft, Ontario, a town that's been on the skids ever since the uranium mines that sustained its economy began shutting down in the 1960s. Now Bancroft has tried to rebound by attracting tourists for mineral-collecting trips. For seven dollars per person, treasure hunters are allowed to dig and search for specimens of hundreds of minerals found in the area. The most common are apatite and feldspar, but garnet and sapphire can also be found. David spent his time there crushing five-hundred-pound rocks with a sledgehammer and testing the remains with a Geiger counter for traces of uranium. The broader purpose of this odd excursion he preferred not to divulge.

In the end, things back in Michigan didn't work out well for David. As was the case with his atomic experiments, his plans for life have been marked by grand and extraordinary ideas that have been plagued by an exotic style of execution. He attended Macomb Community College, though without much enthusiasm, and failed to obtain a degree in applied sciences. As in high school, David worked a series of fast-food and retail jobs, and never earned enough money to move out of his parents' house. And so, after two years in Clinton Township, David decided to reenlist in the navy in mid-2003.

Beyond his planned time in the navy, David isn't sure what he wants to do. All he aspires to, he said during one phone conversation, is "to be happy—like when I was a kid." From someone who had a childhood as difficult and traumatic as his, it's a strange hope.

Then again, perhaps it's not so surprising. David's greatest moments of happiness came when he was alone in the shadowy quiet of the potting shed. There, he sought to solve impenetrable riddles of nuclear physics, while simultaneously seeking, through step-by-step procedures and quantifiable formulas, a sense of control, power, and predictability. That's what he now longs for, as well as a touch of the attention and notoriety, however fleeting and mixed, that his nuclear experiments brought him.

Does David ultimately have the focus and disposition—and support—needed to excel at more conventional forms of scientific inquiry? That's an open question, but given the central role of the atom in his emotional and intellectual life, it's almost certain that David has reached a fork in his nuclear road, not its end point.

Nuclear power, the miraculous force that David Hahn learned about in the *Golden Book* and the Boy Scouts, began making something of a comeback at the dawn of the new millennium, after years of decline. This was partly due to George W. Bush taking office in 2001. He called nuclear power a "major component" of any long-term solution to meeting America's energy needs. Plant operators were applying for extensions of the original forty-year licenses, and engineering firms were promising that safer, cheaper reactors were just around the corner. In the strangest twist, the industry launched a major PR campaign portraying nuclear power as environmentally correct because unlike fossil fuels it contributes little to the greenhouse effect.

Despite all this, popular enthusiasm for nuclear energy remains tepid as a result of accidents such as those at Three Mile Island and Chernobyl. As *Time* magazine observed in the midst of the recent minirevival of nuclear power, "Many homeowners would sooner burn coal in their own fireplaces than live next to a reactor." Furthermore, the industry has yet to come up with a long-term solution to the problem of storing the nuclear waste generated by its power plants, which continues to pile up on-site and at temporary disposal stations around the country.

But the most important obstacle to a major expansion of nuclear power is that it simply hasn't proved to be an economical means of generating electricity. In the late 1970s, when the industry's prospects were rosiest, price hikes by the Organization of Petroleum Exporting Countries had pushed the cost of a barrel of oil up to forty-nine dollars, a figure the Department of Energy predicted would double by the end of the century. Instead, the price of oil began a steady decline and—after bottoming out at twelve dollars a barrel in 1999—still went for about 50 percent less in 2001 than it did twenty years ago.

After growing by 70 percent in the 1970s, global nuclear-energy production climbed by only 5 percent in the 1990s. Now, production levels have begun an absolute fall, with global nuclear capacity expected to drop from a peak of 343 billion watts in 1995 to 172 billion watts by 2020. The last new nuclear reactors ordered in the United States were Palo Verde 1, 2, and 3 in Arizona, and they were commissioned in October 1973, the month that the first Arab oil embargo began. President Nixon envisioned one thousand nuclear reactors dotting the American landscape at the start of 2000. Instead, there were only 104, down from the early 1990s peak of 110.

With fifty-nine nuclear reactors, France occupies second place in the nuclear-power club, followed by Japan with fifty-four, Britain with thirty-five, and Russia with twenty-nine. But as in America, nuclear-power production is declining in relative terms in those countries as well. Meanwhile, Italy has phased out nuclear power, and Belgium, Germany, Holland, and Sweden have decided to follow suit. According to a 2001 report from the International Energy Agency, "Nuclear power is currently being abandoned globally."

The global collapse of the breeder reactor has been even more dramatic than the broader decline of nuclear power. The Japanese government began to operate a small test breeder in 1977 and planned to complete the Monju, a Clinch River–sized demonstration breeder, by the late 1980s. But a series of accidents has largely killed the government's strategy. Most recently, a 1999 calamity at a nuclear reprocessing plant in the village of Tokaimura—about eighty miles northeast of Tokyo— killed two workers and exposed 439 people to radiation. In Russia, a state-sponsored program also stalled after an ambitious start. By 1974, two of the country's experimental breeders experienced sodium fires and explosions so serious that an American satellite is said to have detected them from space. The breeder experienced at least a temporary comeback when Vladimir Putin was elected in 2000. His government announced plans to build sixty-three new nuclear-power plants, of which forty are to be advanced breeder reactors. Given that none of Russia's twenty-nine conventional nuclear-power plants had a complete safety certificate—and many have been cited for hundreds of violations—Putin's program seems wildly impractical.

France poured the greatest amount of money into

developing a commercial breeder program. Rapsodie, a forty-megawatt breeder, began operating there in 1967. Six years later, Phenix, a 250-megawatt breeder, came online, as did the 1,250-megawatt Superphenix in 1986. But none of these reactors performed as advertised. The primary problem has stemmed from constant leaks of the volatile sodium coolant, which breeders use in place of water. At Rapsodie, a sodium spill restricted power production from 1978 until the reactor was closed in 1999. The Phenix plant suffered a string of sodium leaks and was also shut down. (The government claims that it did breed about 15 percent more fuel than it consumed, which would make it the most successful breeder anywhere.) The same fate befell the $4.3 billion Superphenix, which managed to generate only a small amount of energy—it ran for just thirty months out of a twelve-year operating life span—but failed to breed new fuel.

All told, the world's governments have poured more than twenty billion dollars into breeder research and development. All that money has produced no proven technology for "breeding" new fuel, and not a single breeder plant was up and running at the turn of the century.

The grand hope of nuclear-energy advocates today lies not with fission but with *fusion*. Fission generates energy by splitting the atoms of heavy elements like uranium, thorium, and plutonium; fusion seeks to duplicate the sun's source of energy by joining together light, nonradioactive atoms like hydrogen. Back in August 1975, *The New York Times* noted in an editorial a "major breakthrough in fusion research" and predicted that a test reactor could be working "as early as the mid-1980's; commercial applications could become a reality a decade later."

But research into fusion energy has been no more successful than was research into breeder reactors. The primary problem with fusion is that it requires sustained heat of up to one hundred million degrees centigrade, almost seven times the temperature of the core of the sun. That's been difficult to achieve, and attempts to do so have eaten up vast amounts of energy. The high-water mark of research in the United States came in 1994, when a huge doughnut-shaped fusion vessel called a tokamak generated eleven megawatts of power at a Princeton University lab—enough to provide electricity to a few thousand homes but far less power than it took to run the experiment. Today, after worldwide investment of billions of dollars, the most optimistic researchers say that 2050 is the very earliest a fusion-power-producing reactor could be built.

NOTES

his book grew out of a 1998 article that I wrote for
Harper's Magazine. Much of the story told herein is based
on extensive interviews I conducted with the principal
characters, most important David Hahn, beginning in 1998 and
ending in 2003. Important information was also obtained from
police reports and Environmental Protection Agency reports on
David's backyard experiments.

At times, I found David's tale incredible, and there's no way
to be certain of exactly how far he got with his experiments, as
he was the only witness to many of the events. Furthermore,
the events described herein took place some time ago, and
David's recollections—and those of other people I interviewed—
were at times hazy. As should be clear by now, David's model
reactor was just that, a model. It did not and, of course, could
not generate nuclear energy. However, the bottom line, and what

I continually had to remind myself, is that whatever David did, it was scary enough that it required federal and state authorities to bury his laboratory and materials in a low-level radioactive dump site. That's a fairly staggering fact and gives credence to much of what he recounted to me in our numerous interviews.

A number of books proved invaluable in researching background themes to David's story. The best book I found on the history of atomic science is Richard Rhodes's *The Making of the Atomic Bomb*. Mark Hertsgaard's *Nuclear Inc.* offers a detailed look at the history of the nation's commercial nuclear-energy industry. The history of the breeder reactor is told brilliantly by William Lanouette in an April 1983 article in *The Atlantic*.

Catherine Caufield's *Multiple Exposures: Chronicles of the Radiation Age* is a remarkable account of the so-called radium dial women. *Nukespeak*, coauthored by Stephen Hilgartner, Richard Bell, and Rory O'Connor, recounts the hype surrounding atomic science and nuclear energy.

Two books offer a fascinating look at the history of the Boy Scouts: Robert MacDonald's *Sons of the Empire* and Michael Rosenthal's *The Character Factory: Baden-Powell and the Origins of the Boy Scout Movement*.

And of course, a primary reference book is Robert Brent's *The Golden Book of Chemistry Experiments*, which sparked David's interest in science and ultimately led to his attempt to build a model reactor.

CHAPTER 1

20 **One of the studies:** Richard Rhodes, *The Making of the Atomic Bomb* (New York: Simon & Schuster, 1986), pp. 142–43.

CHAPTER 2

31 **As recounted in the delightful:** Stephen Hilgartner, Richard Bell, and Rory O'Connor, *Nukespeak: Nuclear Language, Visions, and Mindset* (San Francisco: Sierra Club Books, 1982), pp. 2–4.

32 **The Radium Chemical Company:** Ibid., pp. 6–7.

34 **By the early 1920s:** Catherine Caufield, *Multiple Exposures: Chronicles of the Radiation Age* (New York: HarperCollins, 1989).

41 **In the flash of an instant:** Rhodes, *Making of the Atomic Bomb*, pp. 713–33.

42 **"First of all let me say":** Ibid., p. 697.

CHAPTER 4

78 **Robert MacDonald, author of:** Robert MacDonald, *Sons of the Empire: The Frontier and the Boy Scout*

Movement, 1890–1918 (Toronto: University of Toronto Press, 1993).

79   **According to <u>Scouting for Boys</u>:** Michael Rosenthal, *The Character Factory: Baden-Powell and the Origins of the Boy Scout Movement* (New York: Pantheon, 1986), pp. 186–90.

87   **An SS major:** Tom Bower, *The Paperclip Conspiracy: The Hunt for the Nazi Scientists* (Boston: Little, Brown, 1987), pp. 110–13.

CHAPTER 5

108   **A New Jersey teenager:** "Random Act Exposes Nuclear Nuisance," *The* (Bergen County) *Record,* December 28, 1997.

CHAPTER 6

119   **By 1948, General Electric:** Mark Hertsgaard, *Nuclear Inc.: The Men and Money Behind Nuclear Energy* (New York: Pantheon Books, 1983), p. 53.

121   **As he recounted, the AEC:** William Lanouette, "Dream Machine: Why the Costly, Dangerous, and Maybe Unworkable Breeder Reactor Lives On," *The Atlantic,* April 1983, p. 35.

123   **The AEC's own figures:** John Fuller, *We Almost Lost Detroit* (New York: Ballantine Books, 1975), p. 2.

125   **According to the book:** Hilgartner et al., *Nukespeak*, pp. 112–13.

127   **A subsequent AEC:** "Nuclear Accident Remains a Mystery," Idaho Falls *Post Register*, July 30, 1995.

127   **In one essay, he:** Karl Grossman, "Captain Plutonium Aims for Food: Radioactive Waste and the Food Supply," *Access* magazine, September 30, 1997.

ABOUT THE AUTHOR

KEN SILVERSTEIN is an investigative reporter for the Washington, D.C., bureau of the *Los Angeles Times*. A former contributing editor to *Harper's Magazine,* in which a portion of this story first appeared, he has written for *Mother Jones, The Nation,* and *Slate,* among others. He lives in Washington, D.C.

ABOUT THE TYPE

The text of this book was set in Filosofia. It was designed in 1996 by Zuzana Licko, who created it for digital typesetting as an interpretation of the sixteenth-century typeface Bodoni. Filosofia, an example of Licko's unusual font designs, has classical proportions with a strong vertical feeling, softened by rounded, droplike serifs. Licko has designed many typefaces and is the cofounder of *Emigre* magazine, where many of them first appeared. Born in Bratislava, Czechoslovakia, Licko came to the United States in 1968. She studied graphic communications at the University of California at Berkeley, graduating in 1984.